Haynes

Pig

Manual

First published in April 2011
Reprinted April 2013
Reprinted in paperback August 2016

British Library Cataloguing in Publication Data
A catalogue record for this book is available from
the British Library.

ISBN 978 1 78521 101 0

Published by Haynes Publishing,
Sparkford, Yeovil, Somerset BA22 7JJ, UK
Tel: 01963 440635
Int. tel: +44 1963 440635
Website: www.haynes.co.uk

Haynes North America Inc.
861 Lawrence Drive, Newbury Park,
California 91320, USA

Printed in Malaysia

**While every effort is taken to ensure the accuracy
of the information given in this book, no liability can
be accepted by the author or publishers for any loss,
damage or injury caused by errors in, or omissions
from the information given.**

Credits

Author:	**Liz Shankland**
Project Manager:	**Louise McIntyre**
Copy editor:	**Ian Heath**
Page design:	**Richard Parsons**
Index:	**Peter Nicholson**
Illustrations:	**Dominic Stickland**
Photography:	**Pages: Liz Shankland (unless stated otherwise).**
	Cover: James Davies

CONTENTS

FOREWORD

Pigs have given me some of the most pleasant, interesting, exciting and frustrating adrenaline rushes of my life, and I hope they'll do the same for you.

Whether you're simply rearing a few weaners for meat, or breeding some wonderful pedigree pigs for conservation or show purposes, it's useful to have a good, solid manual to hand that's based on fact and practical experience – warts and all. A manual just like this one.

Those of us who breed pigs have been through it all: the excitement of the first litter; sorting the piglets, identifying the good breeding stock and weeding out those which are not so good; watching the litter grow; selling those that don't make the breeding standard, or keeping them on and rearing them to pork weight for the freezer.

If you're just starting off it might help you to know that pig keeping doesn't have to be a solitary hobby. Support is available from the British Pig Association (BPA), the custodian of the herd-books of almost all the pedigree pigs presently recorded in Britain, with a membership of 2,000 that's continually growing. There are also many breed clubs, each with a special passion for its own particular breed.

There are agricultural shows too, which are great places to learn about pigs. Showing pigs is itself a very sociable pastime, whether at a local one-day show, of which there are many across the country, or at one of the more prestigious county shows. Showmen and women are very friendly people, always willing to offer help and advice. Just find a show, go along and watch and, once the showing is over, talk to the exhibitors and the judges. They'll be only too happy to answer your questions and point you in the right direction.

So why should you keep pigs? Well, firstly they're fun and companionable. Looking after them will also ensure that you get plenty of fresh air and exercise. In addition, if you rear your own pigs for meat you'll know exactly what's gone into them, so that when you eat them you'll enjoy your pork, sausages, bacon, gammon and black pudding all the more.

It's said that the only thing you can't eat in a pig is its squeak. So if you want to produce your own pork, read this book, get yourself some pigs, and start to enjoy all the benefits of a wonderful pastime.

Alan Rose,
Chairman of the British Pig Association and owner of the Maddaford Herd of pedigree pigs, www.maddaford.co.uk

Introduction

Winston Churchill was absolutely right. Pigs are incredibly intelligent creatures and that's one of the reasons why so many of us are fascinated by them. You may be picking up this book in order to learn about rearing pigs for meat, or you could be thinking of keeping some as pets. Whatever the reason, if you decide to take the plunge you won't be disappointed.

Pigs have personality. There's something about them which appeals to us all. Watch people visiting an agricultural show or a city farm and see how they react when they reach the pigs. I guarantee you'll hear the phrase 'Oh, I *love* pigs!'

Pigs differ from any other type of livestock in that you can actually share an extremely rewarding relationship with them. You're not just the provider of food and bedding; pigs want to interact with you. Inquisitive and intuitive, they will provoke a response in you like no other animal. Like any relationship, it's a two-way thing: pigs love to have their ears and bellies stroked; they'll happily use you as a scratching post, rubbing themselves against you to relieve an otherwise unreachable itch. You, on the other hand, will gain an immense amount of pleasure from providing such services. There's no better way to unwind after a stressful day than stroking a pig.

Above and below: Wild boar have been succesfully crossed with domestic breeds to make easy-to-manage pigs.

In recent years, pig keeping has undergone something of a renaissance. TV chefs like Jamie Oliver, Gordon Ramsay and Hugh Fearnley-Whittingstall have worked hard to convince the viewing public that there's nothing better than home-reared pork. And if that pork has been raised on your own land, by your own fair hands, all the better. Being a pig keeper has never been so fashionable.

But it's not just the resurgence of interest in food quality and traceability that has pushed pigs up the popularity scale; the pet market is growing, too. The past year or so has seen an explosion of interest in the so-called 'micro pig' (see Chapter 1). Knee-high animals, created by cross-breeding one smaller-than-average pig to another, are being snapped up by celebrities like David and Victoria Beckham, Charlotte Church and Paris Hilton, and, it seems, thousands of others are keen to follow suit. These pigs may be bought as pets, but in the eyes of the Department for the Environment, Food, and Rural Affairs (Defra) and the other agencies which oversee the welfare and ownership of livestock, anyone who owns one still has to adhere to the same rules and regulations as farmers and smallholders who make a living out of raising pigs. It doesn't matter whether you own one pig or 1,000, the legal red tape is the same – and so are the husbandry requirements. Pet or not, it's still a pig.

This book is therefore intended as a straightforward, no-nonsense guide to getting started in pig keeping – whether your aim is filling your freezer with pork or getting some charismatic new additions to your family. It's designed to be a one-stop-shop for beginners who need advice on buying and rearing their first pigs, but it should be equally useful to the more experienced pig keeper who occasionally needs to dip into a quick-reference manual for help and advice.

I've made the assumption that, if you're buying a book like this, you'll be thinking of rearing your pigs outdoors – hence the way many of the sections are written. I make no apology for this. It's a basic right of livestock to live in as natural surroundings as possible; while my pigs are with me, they have a wonderful life, and they deserve it.

Keeping pigs is not rocket science, and if you're just raising them to pork weight they can be significantly easier to rear than sheep, goats or cattle. There are always exceptions, but generally few things tend to go wrong in the first six months of life as long as your husbandry is good. However, complications can set in once you venture into the breeding side of things. With this in mind, please don't be tempted to run before you can walk. I often get phone calls and emails from would-be pig keepers who want to jump straight in at the deep end and buy a couple of in-pig sows or gilts. My message is always the same: get to know pigs first. Buy two weaners from a reputable source and watch them grow. Enjoy getting to know them and learning how to satisfy their needs. When you send them for slaughter, get another two of a different breed and see how they compare.

Above: Breeds like the laid-back Oxford sandy and black are a popular choice with smallholders.

Do your homework, talk to experienced pig breeders, weigh up the pros and cons of taking on a much bigger commitment, and make sure you're absolutely certain you can cope with all eventualities before you buy any potential breeding stock. That way, you're far less likely to make mistakes or run into problems. You owe it to your animals.

Liz Shankland

Below: Pigs can be kept on a relatively small acreage.

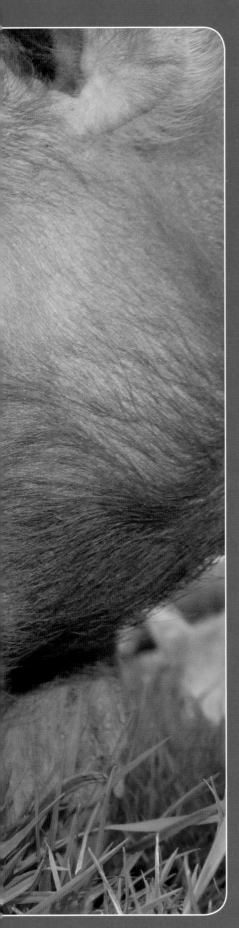

DEVELOPMENT AND BREEDS

History of the breeds

Pigs have been around for a very long time indeed. Research suggests that they existed, in some shape or form, millions of years ago. Eventually, man learned that these wild, forest-dwelling pigs were good to eat (whenever he managed to catch them and not get gored in the process). However, it was around 9,000bc before it dawned on him that he would be better off taking them home alive, fencing them in and taming them – ensuring an easily accessible, continuous supply of these tasty animals. It made perfect sense, and saved an awful lot of hunting time. Consequently, the pig was one of the first animals to be farmed, and it adapted to domestication with relative ease.

All domestic pigs (*Sus scrofa domestica*) are descendants of the wild boar (*Sus scrofa*), a strong, bristly creature with a long snout purpose-built for rooting in the ground and foraging for food. As with all species, domesticated pigs have developed different physical characteristics according to where they were raised, partly due to geographical conditions, and partly due to regional preferences and selective breeding. The wide variety of colours, shapes and sizes which still survive, and which make pigs such a fascinatingly diverse species, can be seen on the following pages.

Foreign influences

For many centuries, pigs raised in the UK retained some similarities with their wild boar ancestors; they were long-legged, slender animals, with long snouts, ideal for rooting about and foraging for food. However, during the latter part of the 18th century small Asian pigs began to be imported and were cross-bred with native stock to improve conformation and productivity. Chinese and Siamese, as well as Neapolitan pigs – shorter and heavier animals with smaller noses and 'dished' faces – were used to increase bulk and improve litter sizes. An additional consequence, of course, was that the appearance of the native breeds changed considerably.

Interestingly, the lively and spirited Tamworth, with its long, rangy body, was not considered worth 'improving', so the breed remained relatively untouched. As a result it remains the closest living relative to the wild boar.

Below: Young wild boar.

Dr Martin Goulding

James Davies

Above: Tamworths are considered the purest of the native breeds.

Economic pressures

Until the early 1930s most of the traditional breeds played an important part in pork production. However, the large white gradually emerged as one of the favourites among commercial producers, with the result that the middle white and the Berkshire suffered a significant decline.

Worse was still to come a decade after the end of the Second World War. In 1955 the UK Government, concerned about the ability of pig farmers to compete on the international market, commissioned research into the development of pig production. The advisory committee, led by Sir Harold Howitt, came back with a devastating blow for all but three breeds. It said that one of the main 'handicaps' facing the British pig industry at the time was the sheer 'diversity' of pigs available! It recommended that 'a single type of pig' should be selected, and went on to suggest that only the large white, the landrace and the Welsh should be used if British farmers were to increase exports and improve profitability.

Not surprisingly, the Howitt Committee's advice sounded the death-knell for many breeds. As commercial hybrids using the 'big three' were created, herds of traditional pigs dwindled. Within the next 15 years, four breeds – the Cumberland, the Dorset gold tip, the Yorkshire blue and white, and the Lincolnshire curly coat – became extinct. Meanwhile, the two varieties of saddleback – the Essex and the Wessex – were amalgamated, eventually becoming the British saddleback.

Changing tastes

As fast-growing, economical breeds came to dominate the commercial industry, consumer tastes also began to change. With shoppers becoming more health-conscious, the demand for leaner meat grew. The result was that cheap, super-lean pork pushed the naturally fatty native breeds into obscurity. By the early 1970s breeding numbers had fallen to frighteningly low levels and were relegated, in the words of the British Pig Association (BPA) to 'curiosities to be displayed at county shows'. To borrow a phrase from Henry Ford, it was a case of 'any colour, as long as it's white'.

But there was a saviour waiting in the wings. In 1973 a handful of breeders, determined to halt the decline of so many important breeds, formed the Rare Breeds Survival Trust. The organisation began raising public awareness of the need to preserve the diversity and genetic heritage of British breeds and, slowly, the native breeds have recovered from the brink of destruction and appear, at last, to be holding their own.

Recent years have seen a resurgence of interest in locally reared produce and food traceability, and there has also been an increase in the use of pigs in conservation – making the most of their natural ability to till the land in order to regenerate overgrown areas for the benefit of nature.

In addition, more and more ordinary people are buying land and raising their own livestock – whether as the first step towards self-sufficiency, or simply to have more control over and confidence in the meat they're eating. More and more of these 'hobby farmers' are choosing to keep pedigree pigs. They may each only have a handful of registered stock in their herds, but if they keep breeding and registering offspring which meet the requirements of the various breed standards, they're helping to keep a variety of bloodlines alive for the future.

Below: Traditional breeds are making a comeback with smallholders.

Pig breeds – a matter of personal preference

The world would be a really boring place if we all liked the same things. Visit any agricultural show and you'll see doting breeders preening their pigs: white pigs, spotty pigs, striped pigs, black pigs, ginger pigs, and even curly-coated pigs.

But why do people favour one particular breed more than another? Choosing a pig is rather like choosing a car: you can go out looking for one for a specific purpose, but appearances can have a big influence on what you end up buying. It may be that you should go for the porcine equivalent of a reliable and easy-to-handle family estate – like a Gloucestershire old spots. However, you could end up with a Tamworth – the Ferrari of the pig world. People may tell you it's too fast and totally impractical, but it will be a real head-turner and great fun to own.

Even if you're rearing pigs purely for meat, there's no reason why you shouldn't have the additional benefit of having something that looks good too. Some call it 'the tingle factor' – that special feeling of excitement you feel when you set eyes on your favourite breed. There's absolutely nothing wrong with having the best of both worlds.

Wendy Scudamore

Above: Kunekune pigs have been described affectionately as the 'Walt Disney cartoon version of a pig'.

Below: An Oxford sandy and black (left) meets a middle white.

Lop-eared or prick-eared?

Beginners are often advised to opt for the lop-eared varieties because a pig whose sight is partly obscured by floppy ears is considered more docile. The logic behind this is that if a pig doesn't have much of a field of vision it won't be able to see where it's going very easily, and so isn't likely to go haring about the place. It's also assumed that a pig which has ears standing up to attention can see where it's going, and therefore will waste no time in racing off to explore. Bear in mind that these are just generalisations; you can have extremely active and difficult-to-manage lop-eared pigs, just the same as you can have quiet and laid-back prick-eared ones.

Above: A large black sow.

Below: A Berkshire piglet.

Above: Sleeping Tamworths.

The right pig for you

There are lots of things to consider when choosing a breed, but don't forget that if you start off simply rearing a couple of weaners for the freezer, it's not a long commitment. You'll normally buy them at eight to ten weeks old and keep them until they reach pork weight at about six months (sooner with breeds which mature earlier). Therefore you're only going to be responsible for them for a maximum of four months – a little longer if you decide to rear them for bacon. So why not start your pig-keeping career by choosing two of one particular breed, then try a different breed next time? See what best suits you – and your palate.

You should, however, bear in mind your location and whether you're going to keep your pigs indoors or out. If you live on top of a mountain and want your pigs outdoors all year round, you'll need some which can withstand the worst of the elements. Some breeds are hardier than others and will thrive in all weathers. Others require more protection from the elements. White-skinned pigs with short coats can burn badly in strong sun; and in winter they don't do as well outdoors, because much of the food they consume goes towards keeping them warm instead of building a meaty carcass.

You also need to think about what you plan to do with your meat. If you want to sell to a butcher who likes a lean carcass with plenty of length and big hams, that rules out many of the old traditional breeds, which naturally carry more fat and are less muscular.

Overview of the breeds

The kind of pigs you'll see in the UK are roughly split into the traditional breeds, the modern or 'improved' varieties, which are popular in commercial systems, and a few imported ones. Traditional breeds are often slow-maturing, and produce fattier meat. Modern breeds, on the other hand, have been improved through selective breeding for rapid growth, leanness and prolific litters. Boars from some of the imported breeds, like the Duroc, the Hampshire and the Piétrain, are more often used as terminal sires (ie they're mated with a pig of a different breed to pass on desirable characteristics) than to breed pure litters.

Berkshire

Colour/markings: Black coat (occasionally dark brown), with white on face, feet and tip of tail
Classed as: 'Vulnerable' by the Rare Breeds Survival Trust (RBST)

The compact size, docile nature and distinctive markings of the Berkshire make it an extremely popular breed for first-time pig keepers. The Berkshire is living proof that not all prick-eared pigs are difficult to handle, and it is often referred to as the 'ladies' pig' because of its easy-to-manage nature. It does well outdoors, withstanding extremes of temperature all year round.

Originally from the Thames Valley, it was once a large tawny red pig with black spots. Although most of the Berkshires you see will have the standard black base colour, some are a deep reddish-brown.

Light-boned, it produces a good ratio of lean meat to fat and can be 'finished' (ie ready for slaughter) early, at around four to five months. Don't be put off by the black coat – the rind on the meat will be white.

The breed is particularly popular in Japan, where it's known as kurobuta ('black pig') and its meat is highly prized.

British landrace

Colour/markings: White
Classed as: 'Endangered' by the RBST

One of the most popular commercial breeds in the UK and other parts of Europe, renowned for its long, lean carcass. First registered in 1896 in Denmark – where it remains the breed of choice – it was introduced to the UK from Scandinavia in 1949 and the British Landrace Pig Society was formed in 1950.

Fast-growing and prolific, the landrace became a favourite in cross-breeding programmes as producers worked to improve the quality of their hybrid offspring. It has been estimated that more than 90% of hybrid gilts

produced in western Europe and North America have landrace blood in their make-up. Sows have large litters and piglets grow fast, making the breed economical to rear.

The breed is docile and easy to manage, making it suitable for intensive systems. However, few landrace are reared outdoors all year round because of their pale, thin skin and fine coat. Lop-eared and white, it's occasionally mistaken for the British lop or the Welsh.

British lop

Colour/markings: White
Classed as: 'Vulnerable' by the RBST

The rarest of all our native breeds, with fewer than 200 registered breeding sows, the British Lop has links to the lookalike Welsh and landrace, as well as the now-extinct Glamorgan, Cumberland and Ulster pigs.

This breed originated around the town of Tavistock in the west of England and was known locally as the Devon lop or Cornish lop. For many years the breed was only seen in areas either side of the Cornwall/Devon border, though occasionally appearing in Somerset and Dorset. In the early 20th century, the breed was known as the 'long white lop-eared pig', but this was eventually shortened to 'British lop' in the 1960s.

A good choice for indoor or outdoor rearing, this breed is hardy enough to withstand cold temperatures, but does need protection from the sun. One of the most docile breeds, it's ideal for beginners.

British saddleback

Colour/markings: Black with white shoulder 'saddle' extending down to the legs

Classed as: 'At risk' by the RBST

An increasingly popular choice with smallholders and one of the most easily recognised breeds. There were originally two types of saddleback – the Essex, which came from East Anglia, and the Wessex, which originated from the New Forest in the south of England.

The Essex and the Wessex were both predominantly black pigs; the Essex had white feet and white on the end of the tail, while the Wessex had a shoulder 'saddle' or belt which continued down over the legs. By the 1960s numbers of both breeds were declining, and by 1967 the two breeds had been merged to create the British saddleback.

Best kept outdoors and allowed to free-range, it's hardy and able to cope with all kinds of weather.

Duroc

Colour/markings: Deep reddish-brown

The breed as we know it today was developed in the USA, but historians are unlikely to ever agree on its true origins. There are suggestions that red pigs might have been imported from the Guinea coast of Africa, from Portugal, or from Spain. The 21st-century Duroc is the result of selective breeding from two different strains of red pig – one from New York and one from New Jersey. The name 'Duroc' might seem to suggest a French connection, but the breed was, in fact, named after a famous thoroughbred stallion

owned by one of the American developers of the breed.

Highly valued as a terminal sire, the breed is one of the most widely used in the world for improving carcass conformation. Consequently little is heard about the quality of meat produced by pure-bred pigs. Those in the know, however, praise the meat above all others, partly because of the high amount of intramuscular fat ('marbling') which improves succulence.

Extremely active and best reared outdoors, the Duroc can be a bit of a handful for beginners and can be dangerous in inexperienced hands.

Gloucestershire old spots

Colour/markings: White with random black spots
Classed as: A 'minority' breed by the RBST

This is the original 'orchard pig' or 'cottagers' pig', a hardy breed traditionally reared on windfall apples and whey. A white pig with black spots, folklore says that the spots were caused by falling apples which bruised the skin.

The county of Gloucestershire was famous for cheese

Richard Lutwyche

production and apple orchards and the GOS converted the waste from both into valuable protein – whey from cheese production and windfall fruit formed the basis of its diet.

The first pedigree records of pigs began in 1885, though the GOS is thought to have existed for more than 200 years. The breed was registered in 1914. Campaigners scored a victory in the summer of 2010 when the European Union granted Protected Food Name status to GOS meat, so that only meat from pedigree stock can be labelled with the breed name.

Hampshire

Colour/markings: Black with shoulder 'saddle'; prick ears

At first glance similar in appearance to the British saddleback, this breed is black with a white 'belt' or 'saddle' around the shoulders and forelegs. Importantly, unlike the British saddleback, it has pricked ears. Originally descended from the Wessex saddleback, the Hampshire was developed

in the USA and is considered by many to be an excellent terminal sire, making it a favourite in commercial cross-breeding programmes. The breed is fast-growing, long-legged, and sows produce large litters. The meat – whether pure bred or crossed – is lean and muscular.

The Hampshire arrived in the UK in 1968 and is now popular worldwide. Although large pigs, they're manageable and can be raised indoors or out.

Richard Lutwyche

Iron Age

Colour/markings: Dark brown to black

Remember the Tamworth Two? Well, the pigs that escaped on their way to an abattoir in Wiltshire in 1998 and stirred worldwide interest weren't actually pure Tamworths. They were Iron Age pigs – Tamworths crossed with wild boar.

Native wild boar disappeared from the UK by the 13th century, and although more were reintroduced during the 1600s these, too, were eventually hunted to extinction. In the 1980s small numbers began to be imported from other parts of Europe to be reared in captivity for their meat. Many ended up escaping – particularly during the 'Great Storm' of October 1987, which destroyed buildings and tore down trees – and began cross-breeding with ordinary pigs. There are now numerous herds living wild in the UK, though few are thought to be as pure as when they escaped.

Pure wild boar are covered by the Dangerous Wild Animals Act, 1976, and stringent security measures must be satisfied before a licence is granted. However, wild boar hybrids – or Iron Age pigs – can be kept just the same as domestic pigs, without any special regulations.

Iron Age pigs have the appearance of wild boar, with thick, bristly coats, long, slender snouts, and straight tails. The offspring are just like wild boarlets, too – born with camouflage stripes which fade as they get older.

Kunekune

Colour/markings: Wide variety of colours, some plain, some spotted. Some are born with goat-like neck tassels – 'piri piris'

Best known in the UK as a pet pig breed, the kunekune (pronounced kooney kooney) can be just as good for meat as many of our other breeds.

The name, quite unflatteringly, means 'fat and rounded' in Maori. The kunekunes in Britain came from New Zealand, but the exact origins of the breed are uncertain. New Zealand had no indigenous livestock, and it's thought that whalers operating in the surrounding waters may have brought them in the 1800s to trade with the Maoris for other goods. The breed could be of Polynesian descent, as there are similar pigs in the South Pacific Islands, as well as in Asia and South America.

Whatever its background, the breed is now popular in many parts of the world. Relatively small – standing between 60 and 75cm high (24–30in), and weighing 54 to 109kg (120–240lb) – the kunekune responds well to training and can be taught to obey commands, just like a dog. The first pigs arrived in Britain in 1992 and the breed has become favoured by smallholders who want small, manageable and docile pigs which prefer to graze rather than root.

Large black

Colour/markings: Completely black
Classed as: 'Vulnerable' by the RBST

A large but docile lop-eared breed renowned for its gentle manner, this is the only completely black pig native to the UK. Its colouring helps it resist sunburn, so it does well outdoors in summer. Often referred to as the Cornish black, the breed's origins are in Devon and Cornwall, although there were some herds of smaller all-black pigs in East Anglia.

The breed is sometimes known as the 'elephant pig', because newborn piglets resemble tiny elephants with their huge ears and straight tails. The hair is fine, soft and silky. Despite the black skin, the meat does not have a black rind; surface pigmentation is removed during the butchery process.

Large blacks are economical pigs to raise, as they're good grazers which convert foraged food very efficiently.

Large white

Colour/markings: White
Classed as: 'At risk' by the RBST

Often described as 'the world's favourite breed', the large white traces its origins back to the old Yorkshire – a large, leggy pig which was cross-bred to produce not only the large white, but the middle white, small white, the Cumberland and the Leicestershire.

It was one of the founder breeds of the National Pig Breeders' Association – now the British Pig Association. It's popular in its own right, but is also favoured for cross-breeding in order to improve carcass quality.

The large white is easily distinguished from other white-skinned pigs like the British landrace, British lop and the Welsh because of its large, erect ears. Despite its prick ears, it's considered a fairly docile, easily managed pig.

Originally developed as a hardy outdoor breed, it became popular with commercial farmers and plays an important part in intensive indoor systems across the world. Between 1970 and 1973 an estimated 8,500 were exported across the world, notably to the USA where the breed is known as the American Yorkshire.

Mangalitza

Colour/markings: Red, blonde or swallow-bellied
(black with a pale underbelly)

A recent arrival in the UK and probably the most unusual of all breeds, the mangalitza has a unique curly coat that comes in three different colours. Found mainly in Austria, Germany, Hungary, Romania and Switzerland, it's a relative newcomer to the UK, having been first imported in 2006.

In 1993 the world population fell to fewer than 150

Dan Walker

Dan Walker

sows, but a rescue mission by a small number of dedicated breeders brought it back from the brink.

The mangalitza contains some traces of a now extinct English breed – the Lincolnshire curly coat, which was last seen in the 1970s. During the 1920s and 1930s hundreds of curly coats were exported to Austria and Hungary for cross-breeding with the native mangalitza.

Known as 'the lard pig' because of its ability to produce vast quantities of rendered fat, the mangalitza has found favour with businesses which specialise in niche market charcuterie products. The meat is well marbled with fat, producing a moist texture, and is useful in long curing processes.

Middle white

Colour/markings: White
Classed as: 'Endangered' by the RBST

Beauty is in the eye of the beholder, and this has been called a 'beautifully ugly pig'. It has a squashed or 'dished' face – a legacy from cross-breeding with Asian pigs – which makes it unmistakeable from any other breed.

The breed was first recognised in 1852 at the Keighley Agricultural Show in West Yorkshire. The judge decided that several sows owned by one exhibitor were too small for the large white class, and too big for the small white class, so a third class was created, specifically for these 'middle' whites. Although the breed originated in the north of England its popularity soon spread south, so much so that it became known as the 'London porker'. It built an international following, too. In Japan, where the pigs are known as 'middle Yorks', the meat is so popular that a statue was erected in appreciation of the breed's 'outstanding eating qualities'.

The patron of the Middle White Pig Breeders' Club is celebrity chef Antony Worrall Thompson, who has become a hard-working ambassador for middle white pork, using it in his own restaurants and promoting it in recipe books.

Oxford sandy and black

Colour/markings: Sandy to deep rust coat with black blotches; pale hair on feet, head blaze and tip of the tail

Classed as: 'At risk' by the RBST

One of the oldest British pig breeds, having been around for almost 300 years, this is often described as the 'plum pudding pig' because of its markings. The colour can be anything from pale sand to deep ginger, with black markings – random blotches rather than spots. The head is slightly 'dished' and the ears are slightly lopped but not over-large. Both the Berkshire and the Tamworth were used to create the Oxford we know today.

The breed is generally regarded as docile and hardy, coping well with extremes of temperature. Oxfords are more likely to graze and forage than to root, which makes them popular with pig keepers who don't want their ground ploughed up too quickly.

The Oxford is a popular choice as a bacon pig, as it can be reared longer without the danger of too much fat being laid down.

Piétrain

Colour/markings: White with black-grey blotches

This breed takes its name from the village of Piétrain in Belgium, where it was created by crossing a variety of English and French breeds. This piebald-looking pig is the most powerfully built of all breeds, with huge, double-muscled hams and an extremely lean carcass. It is to pigs what the Belgian blue is to cattle. The Piétrain is one of the most popular for cross-breeding to improve carcass conformation, but meat from pure-bred pigs is rarely seen for sale.

Its popularity soared after World War Two, but it took until 1964 before the breed arrived in the UK. Despite its redeeming features as a meat pig, the Piétrain had one serious problem: it could drop dead suddenly if stressed. Much work has been done to try and eliminate the Porcine Stress Syndrome gene, and a 'stress-negative' line was developed in the 1980s.

Gillo Pedigree Piétrain Pigs

Gillo Pedigree Piétrain Pigs

Tamworth

Colour/markings: Red (ginger), bristly coat
Classed as: 'Vulnerable' by the RBST

The most primitive-looking of the UK breeds, the Tamworth has a distinctive ginger coat, long snout and prick ears. The Tamworth was considered a poor choice for cross-breeding when Asian pigs started to be introduced into many pigs' bloodlines. It's therefore the purest of the British breeds and the most closely related to the wild boar.

The Tamworth takes its name from the town in Staffordshire. In the early 1800s Sir Robert Peel played

James Davies

a leading role in developing the breed. He had a herd at his home, Drayton Manor, which he cross-bred with pigs known as 'Irish grazers'.

Excellent dual-purpose pigs, famed for both their pork and bacon, Tamworths are among the hardiest of all breeds. They thrive outdoors in all weathers, but because of their lively and inquisitive nature are not suitable for indoor rearing. Their thick, bristly coats help them to resist sunburn, but, like other light-skinned breeds, their ears need some protection.

Despite numbers falling dangerously low in the 1970s, the Tamworth is enjoying something of a comeback, with more than 400 sows currently registered.

Vietnamese pot-bellied

Colour/markings: Black; white with grey patches

Having been developed in Vietnam in the 1960s as a dwarf pig, the breed first found fame in America some 20 years later when large numbers were imported as novelty pets. They were, without doubt, the 1980s equivalent of today's micro pigs (more on these later). Celebrities like actor George Clooney paid considerable money for their snub-nosed pets and the fashion craze spread all over the world. Unfortunately, many people who bought piglets found that they didn't stay cute for long; some displayed aggressive behaviour and many were abandoned after becoming difficult to manage.

Mossburn Animal Centre

Mossburn Animal Centre

The pigs are fairly small – around 42cm (16in) tall – and have a tendency to put on fat if not encouraged to exercise. The original black pot-bellied pig was extremely fat, with short, thick legs. The white is slightly larger and hairier and has a straighter back than the black. It has been bred to have a better temperament than its predecessor. Both varieties farrow only once a year, but have large litters.

Welsh

Colour/markings: White
Classed as: 'At risk' by the RBST

The only indigenous Welsh breed, this is a white pig with fairly long lop ears which almost meet at the nose. It has a long, level body and deep, strong hams.

Various types of white lop-eared pigs had been reared in the counties of Glamorgan, Carmarthen, Pembroke and Cardigan for generations, but it was not until the 20th century that the breed moved to a more formal footing. The Glamorgan Pig Society was formed in 1918, in order to help meet a post-war demand for pork and bacon. Then in 1920

the Welsh Pig Society for West Wales was formed, and in 1922 both of these societies joined forces to form the Welsh Pig Society.

Around 50 years ago the Welsh was one of the most popular breeds in the UK, widely used in commercial herds and cross-breeding. However, its popularity began to wane as shoppers decided they preferred the leaner meat of the cross-bred pigs, with the result that by 2002 there were only 82 registered, and the Welsh was declared a rare breed. Soon afterwards the Pedigree Welsh Pig Society was founded, and successfully attracted funding from the Welsh Assembly Government to help promote the breed. Numbers are gradually rising, with more than 450 sows now registered.

The 'pet' pig market

Tony Price

Above: This young Tamworth boar was taught to walk with a harness as a piglet, but when fully-grown he could weigh more than 400kg and be less easy to manage.

Everyone seems to love pigs, and the past few years have seen a rise in the number of people wanting to keep pigs as pets. Kunekunes and pot-bellied pigs were once the smallest pigs available, but the current craze is for 'micro pigs' – selectively bred mongrels which, breeders say, will grow no taller than knee-high.

The height of a person's knees can vary, of course, so the description is open to interpretation. In 2010 one breeder was sanctioned by the Advertising Standards Authority for claiming her pigs would grow no more than 12–16in (30–41cm), following reports from buyers that theirs had grown way beyond the upper limit. She has now been banned from using this claim in advertisements.

Pictures of young micro piglets sitting in oversized teacups have led to them being erroneously named 'teacup pigs' – and to confusion about the eventual size of the adults.

The truth is that no one can absolutely guarantee the height of any hybrid pig when fully grown. Foundation stock used for most micro pigs is normally Tamworth, Gloucestershire old spots and Vietnamese pot-bellied pigs – and the first two are substantially sized breeds. The micros are created by mating the smallest pigs from litters, but genetic throwbacks are common. A short man and woman may produce a son who towers over them because one of them had tall ancestors – and exactly the same applies to pigs. There can be no guarantees with these hybrids. A further problem is that, because of the practice of breeding 'runt to runt', the parents are not the strongest in the litter and their offspring can suffer a range of health problems.

Celebrity owners have helped make micro pigs the 'must-have' accessory. David Beckham, Charlotte Church, Paris Hilton, and Ulrika Jonsson are among the owners of these creatures, which carry price tags of several hundred pounds. The premium price is for the size, the novelty value and the convenience of having something cute and unusual which won't cost you a fortune to feed.

Micro pigs – just pigs

Owning a micro pig is no different to owning a normal-sized pig when it comes to rules and regulations (see Chapter 2). Unfortunately, far too many people believe they can treat them just like cats and dogs, keeping them indoors as part of the family. And there are numerous unscrupulous dealers who have turned to breeding pigs as a way of making a quick buck – and who will sell a pig to anyone who has the money, regardless of their suitability as owners or the facilities available.

Sadly, even though the whole micro pig phenomenon is a fairly recent one, the classified ad sections are already littered with unwanted 'house' pets looking for new homes. It's a depressing rerun of the pot-bellied pig craze. Owners buying them as sweet, handbag-size animals are horrified when they start growing bigger than expected and displaying normal piggy behaviour – like rooting up carpets and floor tiles, and using their excellent sense of smell to 'forage' for food in cupboards and pet bowls. Although pigs are intelligent animals and can be trained to do many things – including using a litter tray – there's no denying that they smell, particularly the boars.

Several cases of pets becoming aggressive have hit the headlines, notably the 'wedding present' pig bought by former glamour model Katie Price for husband Alex Reid, which developed threatening behaviour at family meal times if he couldn't get more food. Eventually the pig was

Below: This pig's owner wanted a house pet, but it soon destroyed carpets and ripped up parquet flooring.

HOLLY BURNS

Above: The Government booklet with the misleading cover.

rehomed because the couple feared for the safety of their children.

Unfortunately some completely irresponsible dealers are selling uncastrated boar piglets as pets. As discussed in Chapter 3, if you're just rearing boar piglets up to pork weight – to five or six months old – you shouldn't get any problems with aggression or unwanted sexual behaviour, because the pig will not be mature. However, as micro pigs are intended as long-term pets castration is essential. Never buy a boar piglet as a pet if it has not been castrated. Anyone offering one for sale is being totally irresponsible and clearly has no great knowledge – and doesn't care – about pigs.

Having said all this, even castrates and female pigs can become bossy and antagonistic if allowed to do so. Gilts that are not being bred from can exhibit aggressive tendencies when their bodies tell them (every three weeks) that it's time for mating. Vets are reluctant to carry out sterilisations on gilts because the procedure is complex and carries risks, but in one recent case of an uncontrollable gilt terrorising a family in their home every time her hormones started racing, a vet agreed to carry out the operation.

Examples like this should make it clear that pigs are not perfect house pets, and should not be bought with this in mind. Unfortunately, however, the UK Government didn't do much to help deliver this message when it published a document called *Advice for Owners of Pet Pigs and 'Micro' Pigs*, because the cover featured a pig, clearly indoors, standing on a spotless shag pile carpet next to a woman in high heels.

When micro pigs started to attract publicity, Defra issued advice to potential buyers warning them to be aware of the responsibilities and work involved in keeping them: 'Though these animals are typically bought to be kept as pets, in the eyes of the law they remain agricultural animals,' said Dr Nick Coulson, Director of Veterinary and Technical Services for Animal Health. 'This means that they are subject to exactly the same disease controls and regulations as pigs kept in commercial livestock herds. Potential buyers need to understand the full implications of keeping pet pigs, or they could fall foul of the law and put other livestock animals at risk by unwittingly helping the spread of serious animal diseases.'

If you're considering buying micro pigs as pets (and you should have at least two – definitely not one on its own), follow the advice in Chapters 2 and 3. Study the 'Five Freedoms' listed in Chapter 5 and see whether the conditions in which you're proposing to keep your pigs meet the criteria. Pigs need to be able to express natural behaviour, and to deny them this basic right is nothing short of cruel. A pig is a pig, regardless of the label attached to it.

Micro to macro

Margaret Smith, from Ringwood in Hampshire, bought her micro pig, Pigwig, for her daughter Emma (below), because he was 'guaranteed' to stay small. He didn't. By the time he was a year old, he was 26in high, 40in long, and weighed nine stones – and pigs can continue growing for more than two years. He started wrecking the house and eventually had to be relegated to the garden.

9 weeks

Margaret Smith

6 Months

Margaret Smith

1 Year

Solent News and Photo Agency

GETTING STARTED

Have you thought it through?

Taking on any kind of livestock is a big commitment, whether you're embarking on a long-term relationship with your animals or merely raising them for a few months for meat.

More and more people are enjoying their own home-reared pork these days. There really isn't anything better than your own pork, and once you've reared your own you won't go back to the supermarket. However, any would-be pig keeper has to realise that some careful planning is needed before the entertaining little squealers can arrive. Otherwise there may be trouble ahead.

It might help to take a step back for a while and ask yourself the following questions:

Why do I want to own livestock?

Is the aim to produce meat, or do you just want animals in the fields to keep the grass down? Maybe you just want them to enhance the landscape? Remember that there are alternatives to owning livestock; unless you want to raise your own

Above: Deciding to take on pigs is a serious commitment and not one to be taken lightly.

Below: Will all the family be happy to lend a hand if necessary?

Above: Secure fencing is essential with pigs.

animals for meat, you might be better off renting out your fields for grazing. That way you get the benefit of having animals around, while someone else has the responsibility and the expense of looking after them.

Am I sufficiently competent to look after livestock?

Have you the necessary knowledge or training to care for the animals you want? Have you done your homework and found out what husbandry is involved?

Are you physically able to cope with the demands of pig keeping?

Think of where you're planning to keep your pigs and how much fetching and carrying might be involved in keeping them fed and watered. Are you really committed enough to go trudging through muddy fields in all weathers to complete the necessary chores? What would you do in an emergency? Does the local vet treat farm animals? (Find out, because many don't.) And what would happen if you fell ill or had an accident? Is there someone who would be able to step into the breach?

Am I emotionally prepared for keeping livestock?

As the old saying goes, 'Where there's livestock, there's dead stock.' Chances are that you'll have to deal with illness, injury and death. Do you think you could humanely kill an animal if you had to? Something may need putting out of its misery, and no one else may be around to do it.

Do I have the land and buildings that I need?

It's easy to underestimate how much land you'll need, so talk to people with experience before committing yourself. Even if you want your livestock to be completely free-range, you must offer shelter from extreme weather conditions; hot summer sun can be just as bad as harsh winter weather. Similarly, you should have a barn or shed available in case of illness, or for over-wintering stock should the ground

So how difficult is pig keeping?

If you take the simple route and buy a couple of pigs from a reputable breeder, just to fatten for meat, it can be fairly straightforward. When you buy pigs which have just been weaned (separated from their mother at about eight to ten weeks) you'll only have to keep them another four months until they reach pork weight – less if you opt for a modern or a cross-breed, in which case they could mature even sooner.

Any animal that has daily contact with you from weaning age should grow into a friendly, sociable and easy-to-handle creature. Nevertheless, cute little weaners quickly grow into big, strong pigs and, as they're so incredibly food-driven, they can quite happily knock you flying in order to get to the bag of food you're carrying.

You'll have to satisfy yourself that you – or a companion – are strong enough to handle pigs. Even if you're just fattening them for the abattoir, you'll still have to put metal tags in their ears before they go off (more of this later), and that could mean restraining a 70kg animal while you get the job done.

And don't forget that pigs also have teeth and big, strong jaws. They won't necessarily mean to nip you, but they do have a tendency to sniff and nibble at anything out of curiosity, and to check if it's edible. For this reason, visitors, children and dogs should never be allowed in with them unsupervised.

Preparing for your first pigs

✔ **Make sure you comply with the rules and regulations**

✔ **Secure your fences**

✔ **Arrange suitable accommodation**

become too waterlogged or need resting. Don't forget you'll also need somewhere secure to house feed and bedding – it's surprising how much you'll go through.

Are my boundary fences up to scratch?

Never trust to luck. Fences and hedges have to be secure – for the protection of your livestock, to protect neighbours' land and property, and for the safety of passers-by. You should also consider finding an insurance policy that covers not only your livestock but also third-party damage.

Legal requirements

Buying a pig isn't like buying a cat or a dog. There are strict rules and regulations surrounding the raising of livestock and the movement of animals between farms, smallholdings, markets and abattoirs. You first need to register your land as an agricultural holding and inform the appropriate authorities of the livestock you intend keeping.

Procedures are pretty much the same in England, Scotland and Wales, but vary slightly in Northern Ireland and the Republic of Ireland (see Appendix 1 for contact details).

In England, Scotland, and Wales, you need a County Parish Holding (CPH) number, which identifies your land as a holding and allows you to legally keep four-legged livestock or more than 50 birds. In Scotland, a CPH is sometimes referred to as a 'Location Code'. The number will look something like this: 58/421/0086. The first two digits are the county, the next three the parish in which you live, and the final four the actual number of your holding. You'll use this number whenever you buy or sell livestock, move animals on or off your premises, when ordering identification tags, and in various official documents.

Ireland does not use CPH numbers, but you'll need a pig herd number. In Northern Ireland this is obtained from the Department of Agriculture and Rural Development, and in the Republic of Ireland from the Department of Agriculture, Food and the Marine.

Getting registered isn't difficult and it doesn't cost anything. In England, Scotland, and Wales, just contact the relevant Rural Payments agency.

Once you have this number, you'll need to notify your local Animal Health Divisional Office that you have pigs, and you'll be sent a herd number to use if you move the pigs to another location or to slaughter. This number will need to be imprinted on ear tags or slap marks on pigs leaving your holding. When you buy in pigs under 12 months' old, they don't need to be tagged before leaving the vendor's premises – they can move with a temporary identifying mark (eg with stock marker spray). Unlike sheep, they won't need to be tagged unless they leave your land.

Pedigree pigs need additional forms of identification – ear tattoos or notches, depending on the breed – and this is explained in Chapter 8.

Below: Plastic ear tag for breeding stock.

Below: Metal tags are used for pigs bound for the abattoir.

Above: Online movements replaced the old paper forms in April 2012.

Movement licences

Before moving pigs, you must complete an electronic animal movement licence (see www.eaml2.org.uk). Once registered, details of your holding are retained for future movements. The movement must be instigated by the departure premises (i.e. where the pigs are leaving) before it takes place and be confirmed by all parties afterwards. There is a helpline (0844 3358400) to set up movements manually, but there is a charge for calls. Be aware that moving pigs onto your premises means you cannot move any pigs off for 20 days. This is known as the "standstill" period. The only exceptions are if the animals leaving your holding are going straight to slaughter, or if the animals arriving are moved into an approved isolation unit (more of this in Chapter 5). Any other livestock on your land will also be subject to a six day "standstill" at this time.

Below: Pig movements must be recorded.

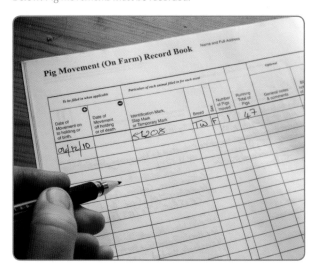

You'll need to record any movements on or off your land – either in a conventional movement records book or on a computer. These records, along with your medical record book (more on this later), have to be kept safe and ready for inspection.

Transportation of pigs

You also need to be aware of the regulations regarding the transportation of livestock. Obviously, the animals have to be fit to travel; those handling the animals must be competent; the vehicle and its loading and unloading facilities must be designed, constructed and maintained to avoid injury and suffering; water, feed and rest (in the case of longer journeys) must be provided; and there must be sufficient space.

There are European Union rules governing the welfare of animals during transport. Anyone transporting livestock more than 65km (40 miles) – or on a journey which lasts more than eight hours – needs to have a certificate of competence. A separate certificate is needed for longer journeys. This means sitting a test at a local college (or at your home for an additional fee) which involves a series of multiple-choice questions regarding the health and welfare of the relevant species.

No farm animals enjoy travelling, so you want to make your pigs as comfortable as possible, with plenty of straw for bedding. They should settle down once you're on the move – so much so that you might have problems getting them out at the other end! If you're planning to transport several pigs from different groups together, have separate compartments in your trailer. Don't give them much feed before a journey, as the pigs may not keep it down, but make sure they've had plenty to drink.

Below: Trailers must be roadworthy.

Organising your pig paddock

How big is big enough?

Pigs don't need a huge amount of space but, as a rule of thumb, don't try and keep more than six per acre. Bear in mind that fields which sit on soft clay and which get waterlogged easily will not last as long as a more free-draining area. It's useful to have a series of small paddocks around which your pigs can be rotated if necessary. Make sure you can provide shelter from strong sunshine – either natural or man-made – to protect your pigs from sunburn and heatstroke. If you can, make a mud wallow too.

Getting fencing right

Good fencing is vitally important when it comes to pigs. You really must be sure that you have an escape-proof enclosure, because if there's the slightest weakness in your fence the pigs will find it. Unfortunately pigs aren't like sheep, which can be contained by well-laid, dense hedges. Pigs are born escapologists, and will challenge any boundary – and win. It's a simple choice: good fencing or wrecked flower beds. Worse still, your pigs might wreak havoc on a neighbour's land, or even wander on to a busy road and put themselves and others at risk.

Traditional stock fencing can be used, but be aware that pigs will often dig under it to reach something growing on the other side. Before long an escape route will have been created and you'll need to get your pig-chasing shoes on. A strand of barbed wire fixed on the bottom can help to discourage curious snouts. Boredom and hunger often spark breakouts, so the more space you can give them to root about the better.

If you can, fix your fence posts into concrete, because pigs rooting around posts and continually leaning against them for a good scratch can destabilise them. If your ground is often waterlogged they may loosen up even quicker, so anchoring them in place will help.

Below: Pigs will try and get under any gap.

Above: Rings in the nose of an Iron Age sow at Cotswold Farm Park.

Ringing

Ringing the noses of pigs to stop them digging is, thankfully, a practice that has largely gone out of fashion – rooting is a natural behaviour which helps pigs forage for food and prevents them getting bored. Ringing can still legally be done to outdoor pigs (but not to those kept indoors), but like castration and tail docking is classed as a 'mutilation' under the Code of Recommendations for the Welfare of Livestock published by Defra (see Chapter 8), and should be avoided wherever possible. It must only be carried out by a veterinary surgeon or another competent person.

Using electric fencing

Electric fencing is often used as an additional precaution, placed inside the perimeter to stop pigs from digging near the fence line. It's also a popular choice as a temporary solution, and in places where permanent fencing isn't wanted.

The beauty of electric fencing – whether powered by mains or battery – is that you can move pigs on to fresh ground when the need arises. Bear in mind, though, that you have to check regularly to make sure it's still working.

Above: Electrified tape used to separate sections of field.

A fence with a flat battery is no good at all. Neither is one that's shorting because the bottom strand is making contact with wet grass.

When using electric fencing as an addition to stock fencing, experts recommend using two strands – one 15cm (6in) off the ground, and the other 23cm (9in) higher.

It's often worth training your pigs to 'respect' an electric boundary by setting up a trial enclosure in a barn before taking them outdoors. Once they get used to the 'hit' of the fence, they should keep their distance. Some pigs, however, never seem to learn.

If you do use electric fencing, make sure you display clear warning signs, to avoid any nasty human accidents.

Fitting gates

If you're able to choose where you site your gates, you can make life much easier for yourself. As well as a day-to-day gate for access, you should also have one that's big enough to drive a trailer through. This will help considerably when it comes to loading time.

Housing: a pig des-res

There's a variety of options when it comes to choosing accommodation for your pigs, but the most popular is the traditional dome-shaped ark. Alternatives include:

- Dome-shaped arks made from wood and galvanised sheeting – probably the most popular option for pigs reared outdoors;
- Barns or stables – whether used as full-time accommodation for indoor-reared pigs or simply to provide shelter as needed;
- Traditional pig sties – low-level stone or brick buildings with indoor sleeping quarters and small outdoor exercise areas, these can be difficult to clean out;
- Triangular-shaped arks, which are quick and simple to make but don't give pigs as much space to move around as conventional ones; and
- Straw shelters, useful as a temporary measure, made by layering straw bales.

Top: An attractive ark at Greenmeadow Community Farm, Cwmbran.

Left: This home-made triangular ark is easy to move.

Below: A plastic calf hutch with added floor for stability.

Dome-shaped arks

This is the best-selling style of pig ark because it offers plenty of space – both for the pigs and for you, when it comes to cleaning out – and because the curved galvanised sheeting makes it weather-resistant and durable. Prices can vary considerably depending on the quality of materials used, but expect to pay anything from £150 to £500, depending on size, quality and features, eg detachable floor, insulation, lifting bars, additional doors and ventilation flaps. A floor is a good investment worth considering if your ground is prone to becoming waterlogged.

Choosing a size

- 1.8 x 1.2m (6 x 4ft) – four weaners to pork weight or two adult kunekunes.
- 2.4 x 1.8m (8 x 6ft) – two 'dry' sows (ie not pregnant nor nursing) or six to eight weaners to pork weight or a sow and litter up to weaning age.
- 2.4 x 2.4m (8 x 8ft) – around 10 to 15 weaners to pork weight or four dry sows.
- 2.4 x 3m (8 x 10ft) – around 20 to 25 weaners to pork weight or six to eight dry sows.

Bear in mind that if you house just a few small pigs in a large ark you'll need more bedding in order to keep them warm. There's nothing like a few spare warm bodies to snuggle up to!

Above: Popular and long-lasting: dome-shaped ark.

Siting your ark

The way you position your pig ark is important. Make sure the entrance faces away from the prevailing wind, and suspend strips of heavy see-through plastic from above the entrance to create a curtain effect, which will help cut down draughts.

Pigs are really clean animals, and tend not to use their sleeping quarters as a toilet, so clearing out your ark shouldn't be too unpleasant. However, you can't teach them to wipe their feet, so in wet, muddy conditions you'll find the bedding needs renewing frequently.

Below: Plastic strips help keep draughts at bay.

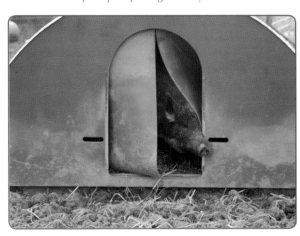

Pig ark in a day

There are many different companies offering ready-made arks or DIY kits, but the basic design is the same and making one isn't that difficult. You can make yourself a superior-quality ark using the best materials and still have change out of £200 – less than half what you might pay for a ready-made ark.

This project will take you step-by-step through the process of making your own dome-shaped ark with a detachable floor. The key to success is getting the timber cut to size for ease of use. It should take less than a day for one person, and will be even quicker if you have a spare pair of hands.

What you'll need

- 6.4m of tanalised smooth-cut timber 100 x 47mm (4 x 2in).
- 21.5m of tanalised smooth-cut timber 74 x 47mm (3 x 2in).
- 25m tanalised rough-cut plank 150 x 25mm (6 x 1in).
- Two sheets of 18mm hardwood ply 2,440 x 1,220mm (8 x 4ft).
- Rolled corrugated steel sheets 2.4m (8ft) wide at the base, 1.2m (4ft) high at the centre. Corrugated steel sheets can be purchased in differing widths. This project uses 2 x 1m (39in). You may need two or three sheets (depending on width) to create the roof with suitable overlaps.
- Woodscrews, 5.0 x 100mm (10 x 4in), 5.0 x 50mm (10 x 2in).
- Metal screws, 5.0 x 50mm (10 x 2in).
- Eight staple-on plates.
- Four 8mm dee shackles.

Preparation

You can save yourself considerable time if you ask your timber merchant to pre-cut the timber for you. Most good suppliers will provide this service at minimal cost. Your timber lengths should be cut and marked as follows:

Using 100 x 47mm (4 x 2in) timber
- 2 x 2.4m lengths (D).
- 1 x 1.6m length (C).

Using 74 x 47mm (3 x 2in) timber
- 2 x 1.6m lengths (B).
- 4 x 1.8m lengths – mark two (E) and two (G).
- 2 x 2.13m lengths (A).
- 1 x 2.2m length (H).
- 2 x 2.26m lengths (F).

Using 150mm x 25mm (6 x 1in) floor planks
- 11 x 2.23m lengths.

Tools

- Electric drill.
- Electric screwdriver.
- Jigsaw.
- Tape measure.

Construction

1 Make the base first. Construct on a firm, flat area. Set out the timber lengths as shown above and fix with 100mm (4in) wood screws. Note that at this stage you're constructing the base face-down.

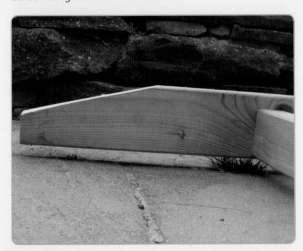

2 Lengths (C) and (D) are the skids for the base. They should be cut at each end as shown above to allow the ark to slide across the ground when it needs to be moved. Lengths (B) are floor supports only. When completed, turn base over so that the skids are now in contact with the ground.

3 Fix floor planks as shown using 50mm (2in) woodscrews.

4 Construct the base for the roof using lengths (E) and (F) as shown above. Note that the front length (F) is upright – unlike the other lengths. This will become the step for the door entrance at a later stage of construction. Fix roof base together with 100mm (4in) woodscrews.

5 Place the roof base over the floor base as shown above and adjust if necessary. The roof base needs to fit comfortably – but not tightly – over the floor base. If the roof base is fitted too tightly then the roof of the ark when completed will be difficult to remove for cleaning.

7 The front ark door should be 610mm (2ft) wide and 915mm (3ft) tall. The door can be square or rounded at the top as required. Make a template, shaped to the design of the door you require, and position it centrally as shown above. Ensure that the bottom of the template is positioned 47mm above the bottom edge of the sheet to allow for the entrance step as constructed in step 4. Mark the outline of the door in pencil and cut out using the jigsaw.

6 The next task is to make the front and back walls using the 18mm hardwood ply sheets. To cut the curve of the ark, place the plywood sheet on a flat surface. Mark the centre spot of the sheet at the top and the bottom. Attach a piece of string to the bottom mark and tie a pencil to the end of the string at the top mark. Keeping the string taut will allow you to draw a semicircular ark shape as shown above. Then using the jigsaw, cut along the pencil line to shape the ark wall. Repeat this procedure to shape the back wall.

8 The front and back walls are fitted to the roof base and held in place by the two supports (G) for the corrugated steel roof. The front and back walls should be screwed to the roof base using the 50mm (2in) woodscrews. The roof supports should be fitted using the 100mm (4in) woodscrews. The supports should be approximately 915mm (3ft) from the ark floor but 50mm (2in) below the top edge of the front and back walls to allow the grooves of the corrugated sheet to sit flush on the ark.

9 The timber length marked (H) is the top roof support and is fitted flush to the top of the front and rear walls as shown above. The overlap at either end allows this length of wood to be used to lift the roof off the base. Note how the support is recessed into the top of the front and back walls and supported on the rear wall by a wooden offcut. A similar support needs to be fitted to the front wall. At this stage, for extra protection, all wood could be treated with an additional coat of wood preservative before the corrugated steel roof is fitted.

10 Fit the corrugated sheeting. Adjust the sheeting to ensure it overlaps in the centre and fits over the three roof supports. Fix sheeting to roof base and the roof supports using 50mm (2in) metal screws. The roof is fixed to the floor base at four points where the outer floor skids are in line with the front and back walls. Above shows the fittings fixed by 50mm (2in) screws. The upper and lower staple-on plates are connected together by the 8mm dee shackles.

BUYING YOUR PIGS

Buying your pigs

If you've got this far through the book, you're pretty determined you're going to get pigs! You may have decided that you want to produce your own pork, or that you want pigs as pets. You've probably also decided the breed you want and know how many you'd like.

But now for the bad news: weaners don't just sit on a shelf all year round, waiting for people to turn up and buy them. The best breeders have waiting lists for their stock, so prepare yourself for disappointment, and accept that you may have to wait a little longer to get what you really want. Buying pigs isn't like popping into one of those pet superstores to pick up a couple of hamsters. Depending on where you live and what breed you fancy, you may find that there are few breeders within easy reach. You may have to travel an hour or so – maybe more – to find the breed you're looking for.

Most beginners want rare breed pigs, which is great, because the traditional native breeds need saving. However, there are reasons why some breeds are rare. As mentioned in Chapter 1, several breeds lost the battle against more cost-effective modern types; others dwindled to the brink of extinction, and it has taken decades of hard work to save them from disappearing altogether.

Traditional breeds aren't popular with intensive pig farms because they tend to have much smaller litters than modern breeds or popular crosses – often as few as six or eight, compared with the high teens or more which are common in commercial hybrids.

So the moral of the tale is, start making contact with breeders as early as possible and place an order. Otherwise, be prepared to compromise and choose a different breed.

Below: Buyers browse before the start of an auction.

Finding a good breeder

If you really have your heart set on a particular breed, start with the breed club – most of which have websites with sections where members post 'for sale' adverts. The British Pig Association (www.britishpigs.org.uk) has a searchable database on which you can seek for information on the 14 breeds it represents throughout the UK, and a section where you can find details of what they have for sale.

Agricultural shows are a 'shop window' for breeders and a great opportunity to meet experts face to face. Most breeders will be willing to talk endlessly about their pigs and answer your questions, so make the most of the opportunity. Expect to pay around £50 for a non-registered weaner from a pedigree herd, and less for cross-breeds. Breeding stock that has been registered as pedigree will be more, of course – a gilt at eight to ten weeks is likely to cost around £100, or possibly more depending on quality and bloodline. The older you leave it, the more the price will go up – understandably, as the breeder is having to feed the pig.

Above and below: Agricultural shows are great places to see different breeds and talk to the breeders.

Other places to buy

Small ads

You could turn to the classified ads section in your local newspaper or online. You may find someone living fairly close to you, and you might even pick up a bargain. But it's not recommended. For every extremely reputable breeder who places a small ad, there are many more who are only in it for the money.

Grabbing a bargain is great, but just because something is cheap it doesn't mean that it's worth the money. So when you see an ad offering pigs at a cheaper price than the one you've been quoted by a reputable breeder, just wonder why. To a good, responsible breeder, the health of his or her herd is paramount. Breeding stock will be carefully chosen and vaccinated on a regular basis and accommodation will be kept clean to minimise infection. No breeder will compromise the health of a valuable herd by skimping on husbandry.

So if you're tempted by newspaper or online ads, just be cautious, and have a good look before buying. As well as looking at the weaners, scrutinise their living conditions, ask to see other pigs in the herd, and make sure you feel comfortable before handing over the money and loading the trailer.

Buying healthy pigs

No matter where you go to buy your pigs, do the same health checks. Take a good look at the animals and make sure they're fit and healthy. Never buy anything that looks too quiet, or generally under the weather, thinking it will be much better when you get it home. The main things to be aware of include:

- General posture and appearance – listlessness, dull eyes, droopy head, dry muzzle, reluctance to stand.
- Discharge from eyes or nose; scouring (ie diarrhoea).
- Generally poor coat/skin – dull or scruffy hair, bald spots, scabs; animal scratching a lot.
- Coughing or sneezing.
- Lameness or other signs suggesting that the animal is in pain.

IF IN DOUBT, DON'T BUY.

Below: Never buy anything that looks less than perfect.

Auctions

Buying at auction is an even bigger minefield for the inexperienced customer. It's great fun and extremely exciting, but full of pitfalls for the novice buyer. You have to know exactly what you're proposing to buy, and be able to carry out all the basic health checks before bidding. You should also talk to the vendor and ask questions about any veterinary treatments which might have been given. Have the weaners been wormed, for instance? Have they been vaccinated?

The big problem at a market is the risk of infection, with so many animals coming and going. You don't know what you might be taking home along with your purchase. When you buy direct from an established breeder, you should be able to ring that person for after-sales advice. Good breeders are normally happy to do this. Buying at an auction is an entirely different way of shopping. As the saying goes, 'You pay your money and you take your choice.'

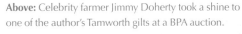

Above: Celebrity farmer Jimmy Doherty took a shine to one of the author's Tamworth gilts at a BPA auction.

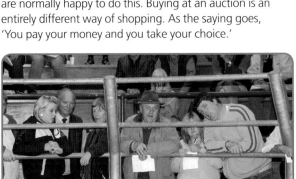

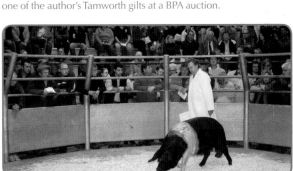

Right and above: The annual BPA show and sale is an excellent place to buy good pedigree stock.

Boars or gilts?

If you're buying pigs to raise for meat, it should make little or no difference whether you buy males or females. Good breeders will often advise you to start with boar piglets because you won't be tempted to hang on to them in the same way as you might gilts.

There really is no difference in meat quality and, if anything, boars will grow better and give you a leaner carcass. You'll be sending them off to the abattoir for pork before they mature and get sexually active, so there's no need to worry about aggressive behaviour, and there's absolutely no difference in the way they need to be handled.

If you're rearing for meat, the only reason to opt for gilts is if you want to produce bacon, which means keeping pigs for eight to ten months.

Boar taint

Newcomers are often reluctant to take on boar weaners unless they've been castrated because they've heard butchers' stories about 'boar taint'. This is an unpleasant smell and taste that affects the meat of some male pigs, but most often older males which have been reared intensively. Traditional breeds don't seem to be as affected, possibly because they're slower to mature and aren't fed diets designed to produce quick, cheap pork.

Boar taint is caused by two compounds in the fat – androstenone and skatole. Androstenone is a steroid, and is produced as the boar reaches puberty. Its function is to act as a sex pheromone that helps make the female pig receptive. Skatole is produced when the amino acid tryptophan is broken down in the gut. It's found in both male and female pigs, but accumulates more in males. Diet and environment can influence how skatole causes boar taint. High levels of fermentable carbohydrate in the intestines is said to help, and the feeding of sugar beet pulp or lupins in the week before slaughter has been suggested. Cleanliness has been shown to be a factor in reducing skatole levels. Skatole is highly concentrated in faeces, and researchers in Denmark, Australia and New Zealand have found high levels in pigs kept in dirty conditions.

A vaccine against boar taint has been produced and has been welcomed by commercial producers who find that surgical castration makes for fatter boars.

At present in the UK, piglets up to seven days old can be castrated whilst conscious and without pain relief. However, there is growing support in the European Union for a complete ban on castration.

Below: Boar weaners are absolutely fine for meat, as long as they are slaughtered before they mature.

Buying for breeding

This opens up a whole new collection of things to look for, and is dealt with in detail in Chapter 8. A pig that's being sold as suitable for raising meat will often be half the price of a pig suitable for breeding, but don't be tempted to cut corners. There will be a reason why the pig has been relegated to 'meat pig' status by its owner. It may have physical features or deformities which should not be passed on to offspring, but which cause no problems if it's just being fattened for the freezer. Pedigree breeding imposes far more restrictions on what should and should not be bred from and, again, this is explained in Chapter 8.

Right: An Oxford sandy & black piglet from a pedigree litter.

Buying pet pigs

As explained in Chapter 1, pigs don't make the best house pets, so if you're thinking of keeping some as pets they must be allowed access to the outdoors – at the very least during the daytime. Also, pigs are herd animals and need the company of their own kind, so don't just buy one and think it will be happy with your dog or cat for company. Think of what's best for the pig.

No one should take pig keeping lightly, so please read this book thoroughly and see if you're ready for what lies ahead before you make a decision.

Margaret Smith

Taking your pigs home

CHECKLIST
- eAML2 movement set up in advance (see p33).
- Suitable transport.
- Straw for the journey.
- Clothes to change into, should yours become messy.
- Money to pay the vendor!

Paperwork

As mentioned earlier, you will need to complete an online animal movement licence before you can take your pigs away, so register on the www.eaml2.org.uk site in advance and give the seller your full name, address and CPH number so that he/she can find you on the site.

Like the old paper forms, the online movement licence includes the name, address, and holding number of the departure premises and contact details for the animals' keeper; the number of animals being moved and their identification marks; information about the vehicle and haulier; and details of the destination premises. The breeder will fill in his or her details as well as yours, and will confirm the movement once the pigs have left. You will also need to confirm the movement has been completed when you receive the pigs.

The new system allows you to select the type of movement required, whether farm to farm, farm to abattoir, farm to show, etc.

When you get home, you must also record the arrival of your new pigs in a movement record book.

Identification

As mentioned in Chapter 2, pigs younger than 12 months old do not need to be tagged before moving. They can be given a temporary spray paint mark, which is noted on the movement licence, e.g. 'green stripe'; 'blue spot'. Pigs over a year old must be tagged with the herd number of the holding of birth before departure.

Transport

It should go without saying that vehicles for transporting livestock should be fit for the purpose, safe, secure, free from hazards and well ventilated. Driving home with a couple of

Below: Straw on the trailer ramp will help with loading larger pigs.

Right: The easiest way to move a weaner.

Mixing litters

You may find your weaners come from different litters that haven't been reared together. This isn't a major problem if they're roughly the same age and size, but be prepared for a bit of initial fighting as the youngsters sort out who is boss. It can look quite nasty when a scrap breaks out, and you'll undoubtedly see scratches and bleeding ears, but they'll settle down eventually. When they arrive at their new home, feed them far enough apart so that they don't have to jostle for position. By morning, it'll be like they had always been together.

Releasing two small pigs into a big field can be a bit disconcerting for them, having been used to being part of a big litter in an enclosed space, so it's worth sectioning off an area of 1.5 x 1.5m (5 x 5ft), just for the first night. This gives them the chance to get used to being in a new place and to get to know you. Sheep hurdles can be used for this, but tie them together securely. Gradually increase the area as they get a little bolder, day by day.

It's important at this stage that your weaners learn to trust you as the new head of their herd. The more attention you can give them, the easier they'll be to handle. Step in with them and sit on the floor. Don't reach out at them – let them come to you, have a good sniff and even a little nibble at your clothes and your shoes. It won't be long until they'll be greeting you like a long-lost friend and heralding your arrival with an ear-splitting welcoming chorus.

frightened piglets scampering around on the back seat of the family car just isn't an option; they're a danger to you and to themselves. They should be transported in a trailer or a vehicle with a section specially designed for livestock. Making a space small and snug (eg by using straw bales to create a more intimate area or by putting them in a large dog crate) will help settle your new charges and encourage them to cuddle up and sleep. You'll probably find that once they get used to the motion of the vehicle, they'll snooze away until they reach the other end.

Good breeders will normally worm weaners before they leave, but check to see if this has been done. Not all breeders do the same: some worm, some don't; some vaccinate against a variety of things, others don't. So ask whether any veterinary products have been administered, and whether any follow-up doses are required.

Loading your pigs

Lifting weaners isn't the easiest – nor the most enjoyable – thing to do. They hate being handled and will scream the place down with their deafeningly loud squeals; they'll wriggle and twist so much you'll have a tough job holding on.

Piglets are born with a kind of self-defence reaction which means they go bonkers and yell as loud as they can when you grab them around the middle. It's a way of alerting Mum that they're being sat or stood upon. Some breeds are worse than others, but no piglets enjoy being held for long. Instead, the tried and tested way of lifting one with a minimum of stress – to the piglet and to you – is to grab it by the back legs and hoist it up.

You must make sure the legs are facing away from you, otherwise you could end up with badly bruised shins. Weaners can be quite heavy at eight weeks, so make sure that the trailer is backed up as close as possible and that you have someone to open and close the loading gates and help with supporting each pig as you lift it on board. One alternative is the 'wheelbarrow' method, with the piglet walking in 'two-wheel-drive' mode towards the trailer while the handler holds the back legs.

FEEDING

Feeding

Pigs are masters of persuasion and will always tell you that they want more to eat. You can guarantee that if you plant too many vegetables in the garden, nothing will go to waste.

For generations pigs were fed anything and everything that was left over at mealtimes. Everyone who kept pigs had a 'swill bucket' in the kitchen to collect any scraps rinsed off plates. However, this is no longer possible in the UK. The foot and mouth outbreak of 2001 was traced back to a farm where pigs had been fed unprocessed waste food. The epidemic that followed saw more than six million animals slaughtered. Exports were suspended, and the total cost of the outbreak – to the farming industry, to the public sector, and to tourism and the rural economy – was estimated at £9 billion.

The subsequent ban on feeding catering waste which was imposed in 2001 was later reinforced by the Animal By-Products Regulations 2005, which specified that this meant 'all waste food, including used cooking oil originating in restaurants, catering facilities and kitchens, including central kitchens and household kitchens'. In short, nothing from your kitchen – nor from anyone else's – can be fed to pigs. Anyone in breach could face either a fine or even imprisonment.

So, in a nutshell, it's now illegal to feed to farm animals anything that passes through a kitchen, whether from catering establishments or residential properties, even if the food prepared is vegetarian. This decision may seem a little draconian, but a total ban does, at least, remove any misunderstandings and grey areas.

There are nevertheless a few exceptions, including waste milk and milk products. However, both you and the person supplying the products have to register with Defra or the equivalent regional agency. More information is available from the Defra helpline (08459 335577) or their website at www.defra.gov.uk/foodfarm/byproducts/guidance/register.htm.

You can source fruit and vegetables from non-catering premises, but this can only be from premises that either don't handle those materials which cannot be fed to pigs, or who have Hazard Analysis Critical Control Point (HACCP) procedures in place to ensure that what's being provided is kept completely separate from banned materials. If in doubt, contact your local Animal Health Divisional Office for clarification.

Below: A variety of containers can be adapted to hold pig feed.

Keeping it simple

By far the easiest way to feed your pigs is to give them a commercial ration that has been carefully blended to contain all the nutrients they need to grow well. Pelleted food – commonly known as pig 'nuts' or 'rolls' – contain cereal crops with added vitamins and minerals. They're formulated with specific stages of the growing phase in mind, from weaner to 'finishing' (slaughter) age.

Some people mix their own feeds, often a combination of traditional cereals like barley, wheat, maize and oats, with added vitamins and minerals for good measure. However, unless you have the time, knowledge and plenty of dry, vermin-proof storage space, it's worth sticking to bagged feed – at least until you've had time to do your own research.

The main objective at this stage, when you're just starting off, is to raise healthy pigs that grow well and produce great pork.

Feeding ready-made pellets is no hassle, the bags are a manageable size, and the feed stays fresh until you're ready to use it.

As with most animals, young, growing pigs require a higher level of protein in their food, and the level should be decreased as they get older if they're being raised for

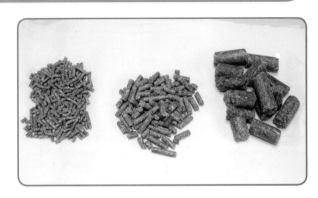

Above: Creep feed for very young piglets (left); grower pellets (middle); sow rolls (right).

Below: Fresh fruit and vegetables are always a welcome addition.

meat. Everyone has their own way of feeding, but here is a typical daily feeding plan for rearing traditional, outdoor pigs up to pork weight. Modern breeds, selectively bred to grow faster, will reach slaughter weight up to two months earlier.

Rearing for pork

Approximate age	Food per day*
8 weeks	2lb / 0.9kg
12 weeks	4lb / 1.8kg
16 weeks	4.5lb / 2.04kg
20 weeks	4lb / 1.8kg
24 weeks	4lb / 1.8kg

*This is split into two feeds.

In this example, you'll see that pigs reared for meat have their feed reduced to 4lb (1.8kg) at 20 weeks – around the time when they start to lay down fat. They are kept on this maintenance diet until slaughter at 24 weeks, to regulate their weight. If they were being reared for bacon, they would continue on the same diet until at least 32 weeks.

Certain factors affect growth: active pigs use more energy than docile breeds; in very cold weather, feed a little extra because they need additional 'fuel' to keep warm. And don't forget the old farmers' rule of 'feeding by eye' – if your pigs look too thin, they need more food; if they look too fat, cut back a little.

Below: Unlike other breeds, kunekunes are uniquely adapted to obtain much of their nutrition through grazing. They need little supplementary feed if grazing is good.

Katie Sully

Above: Tamworth weaners feeding from a trough.

It's good practice to divide the daily feed ration into two portions, to avoid bloating and scouring (diarrhoea). It also gives the pigs something to look forward to twice a day, and means that you have more chance of spotting changes in behaviour and potential problems.

Some people scatter the food evenly across the ground, so that all the pigs have a good chance of getting all they need to eat. This is fine as long as the ground is dry. In wet weather, when the ground gets muddy, you might need a trough. If you do use a trough, it needs to be big enough to allow all your pigs to feed at the same time. Otherwise fights can break out and the less forceful pigs will be pushed out of the way.

Pigs need different amounts of feed at different stages of life. For information on feeding breeding sows and boars, see Chapter 8.

By the time your weaners leave their mother they'll have got used to eating solid food. From around two to three weeks old the piglets will have started showing an interest in their mother's food, and at around the same time the breeder may have introduced a 'creep' feed – a fine, pelleted food (similar in appearance to layers' pellets) which has a protein content much higher than the mother's ration. Sow nuts or rolls are normally around 16% protein, whereas creep feed can be as high as 25%, depending on the manufacturer.

When you get your weaners they may be on grower pellets, which have a slightly lower protein content (though still higher than normal sow nuts). This food can be fed up to pork weight. You may also see a 'finisher' feed advertised. This can be fed towards the end of the rearing process, from about 16 weeks, and is supposedly formulated to stop the pigs laying down too much fat.

However, there's nothing at all wrong with feeding your weaners sow nuts or rolls. Some breeders never feed a grower ration, giving sow nuts right the way through the rearing process. Grower rations also cost more, and unless you're intent on finishing pigs quickly they may not be worth the money.

Whatever you choose to feed, find out what the breeder has been feeding, ask to buy a bag or two, and blend it in gradually diminishing amounts with the feed you want to get them used to. This will help to avoid stomach upsets caused by a sudden change of diet.

Ready-made feed

As I've already said, when you're rearing your first weaners you can't go wrong buying in commercially prepared food. It's been meticulously formulated to provide everything a growing pig could possibly need, it takes the guesswork out of what to feed, and it comes in easy-to-manage quantities. However, if you want to mix your own feeds, that's your choice.

The essential thing to remember is that pigs need to be fed a balanced diet. Everyone is keen to shave a little off the monthly feed bill, but you can't cut too many corners. You may know someone with a market stall or a greengrocer's shop who has promised you all his unsold produce. That's all well and good, but fruit and veg alone won't grow a healthy pig. In the wild pigs are omnivorous and would be eating a varied diet which would include not only vegetation but small animals, providing an important source of protein. Rearing pigs in a captive environment means that natural sources of animal protein must be replaced with something else – hence the inclusion of a variety of nutritious cereals in compound feeds. And don't forget the importance of the essential vitamins and minerals that you see on the

Above: Commercial cross-breeds like these will reach pork weight quicker than traditional breeds and will stay leaner.

paper labels attached to your feed sacks. If you think you can replicate that careful balance by throwing your pigs a couple of boxes of rotting veg a day, you're going to end up disappointed – and so will your pigs.

Below: Weaners can be fed specially-formulated grower rations, or be fed on sow nuts or rolls right the way through.

The golden rules of feeding

The general appearance of your pigs should tell you whether you're feeding them enough of the right foods. They should appear healthy and alert and they should be putting on weight steadily. Fruit and veg can be fed pretty much ad-lib, so having a plentiful supply at your disposal can be extremely useful in keeping them contented. Pigs also eat through boredom, so additional bits and pieces can keep them gainfully occupied. Remember to keep an eye on their faeces, as overfeeding can cause diarrhoea, so if the faeces look too runny you may need to cut back. However, bear in mind that looseness of the bowels can also be a sign of other health problems. If in doubt, consult your vet.

Our ancestors thought that pigs could survive almost solely on potatoes. Now that we know more about the way pigs work, we understand that potatoes are far from a complete food. Furthermore, although potatoes can be fed to pigs, they contain too much starch for easy digestion. This can be overcome by cooking them – but you have to consider the regulations about food which has passed through a kitchen. Maybe a purpose-built vegetable boiler in the garden, just for cooking pig feed, is the answer.

Be extremely careful about the condition of the potatoes you feed. Don't feed any green ones, nor ones which are starting to sprout, because these can cause alkaloid poisoning.

Green, leafy vegetables like cabbage, kale and broccoli are full of nutrition, but not all pigs like them. As with potatoes, the stems can be tough to digest without boiling.

Most pigs love root vegetables, so grow as many different varieties as you can. Carrots always go down well, as do turnips, beetroot and swede. Fodder beet is a particular favourite. It's not just the sweetness that appeals; pigs seem to relish the challenge of steadying the little rugby ball-shaped vegetable and gnawing at it while it keeps slipping away. Don't feed the leaves, however, as these can irritate the lining of the stomach. Similarly, be careful of feeding parsnips, which can cause mouth ulcers, and too many can lead to pregnancy problems and even abortion.

Squashes are always popular, from butternut to giant pumpkins. As with fodder beet, they can be great for keeping pigs entertained if you throw them in whole.

Tomatoes are another great favourite. Pigs go mad for them. Cherry tomatoes can be extremely useful when trying to con your pigs into taking medication; a soluble aspirin (only on veterinary advice) smuggled inside one can soothe many porcine aches and pains.

Your pigs will love most fruits – apart, perhaps, from citrus varieties. Everyone knows that they love apples, but they also adore strong-scented fruits like strawberries, raspberries, bananas and peaches. At the end of the summer, trawl the garden and hedgerows for blackberries, raspberries, sloes and hazel nuts to mix in with your pigs' feed and add interest.

Below: Prepared feeds take the guesswork away from feeding.

Water

Never underestimate the importance of water. You'll be surprised how much a pig can drink during a day – often as much as eight litres for an adult, and twice that amount for a lactating sow. Lack of water can cause salt poisoning (see Chapter 5), which can be fatal.

It's worth investing in self-filling plastic drinkers. These work on the same principle as a toilet cistern, keeping the water level constant. Once connected to the water supply

Above: Self-filling drinkers are useful.

they just need occasional cleaning out – unless, of course, the pigs manage to disconnect the pipe, knock over the drinking bath, or wreck it completely by trying to climb inside! With weaners, make sure that whatever you use as a water container isn't so deep that they can fall inside and drown. Shallow containers are best until the weaners are big and strong enough to be given access to the automatic drinkers.

Above: Make sure piglets can reach the water.

Below: Pigs are drawn to natural water sources.

HEALTH AND WELFARE

Caring for your pigs

Whatever livestock you choose to keep, you have both a moral and legal obligation to care for their needs. The health and well-being of your animals must be of paramount importance. Regardless of how you feel, however busy you are, no matter what the weather is doing when you get up on a particular day, the pigs must be looked after. Similarly, if they need veterinary attention or medication you must never neglect your duties.

If you can't manage something yourself, or are unsure of what's wrong, *call the vet*. Don't just ignore the problem and hope it will go away.

Being a responsible owner means granting your animals the 'Five Freedoms' – a list of basic rights originally drawn up by the Farm Animal Welfare Council in the UK, but now internationally accepted:

1 Freedom from hunger and thirst – Animals must have ready access to fresh water and a diet to maintain full health and vigour.

2 Freedom from discomfort – They should have an appropriate environment, including shelter and a comfortable resting area.

4 Freedom to express normal behaviour – There should be sufficient space, proper facilities, and animals should have the company of others of the same species.

5 Freedom from fear and distress – Mental suffering must be prevented by ensuring animals live in appropriate conditions and are given appropriate treatment.

The Defra website (www.defra.gov.uk) has excellent guides to the welfare of all farm animals. Look in the 'Animal Health and Welfare' section, then click on 'Animal Welfare', then 'On-Farm Animal Welfare'. Alternatively phone them on 08459 335577 to request hard copies. In addition the UK Government website at www.direct.gov.uk has a guide for new pig keepers in its 'Environment and greener living' section.

3 Freedom from pain, injury or disease – Appropriate care should be given to prevent health problems, and there should be rapid diagnosis and treatment.

Do your homework

There's no substitute for getting some experience with pigs before you take on some of your own. Try and seek out a local pig breeder who will let you spend some time observing and helping out. It can be a lot easier mastering tasks like injecting if you can do it alongside someone with experience. If you can't find a friendly farmer, try to find a community farm that has pigs, or seek out a specialist training course. However, check the credentials of the course tutors, find out what will be covered, and establish how much 'hands-on' experience you'll get before handing over any money.

Finding a vet

Don't wait until your pigs fall ill before registering with a vet. Vets who treat livestock are hard to find, although there appears to be no shortage of those who treat small animals. You may have to look outside your immediate area and possibly pay an additional call-out charge to cover the extra travelling involved.

Arrange a visit before you need help – simply to check over your livestock, make sure everything is okay, and to draw up a calendar of routine tasks to be carried out, such as worming and vaccination. The vet will be happy to show you how to administer drugs and explain, for instance, the sites for different types of injections.

You may feel happier calling out the vet even for routine jobs the first couple of times – and you should always do so if you don't feel sufficiently confident. By the time subsequent treatments are needed you may feel able to do the job yourself.

By law, you have to keep a record of all medicines administered to your livestock. This should include the date of purchase, the name of the drug and quantity, the supplier, ID marks of the animals treated, the dates treatment was given, the date the withdrawal period ended, and the name of the person who administered the medicine. Records can be written or computer-based, but they have to be kept up to date and available for inspection if required. You can buy

Above: Bob Stevenson, the BPA's consultant vet, examines Berkshires on a routine visit to Greenmeadow Community Farm in Cwmbran, Monmouthshire.

computer softwear for keeping farm records, or start an online record on the Agriculture and Horticulture Development Board website (www.ahdb.org.uk).

Below: Remember to note details of any veterinary products you administer to your pigs.

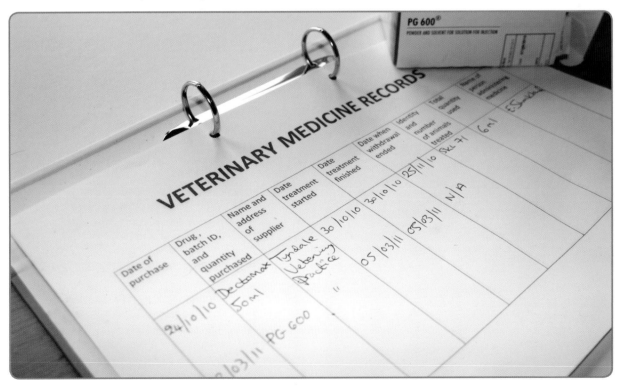

Isolation units

If you already have pigs, bringing more on to your farm is a high-risk activity. New pigs bought from market or from another breeder can be carrying infections and diseases – as can your own pigs when they've spent time on someone else's premises.

It's good common sense to isolate any pigs you buy until you're happy they're healthy. Many breeders do this as a matter of course, rather than risk the health of the entire herd. For breeders who have hire boars to visit on a regular basis, or whose sows and gilts go off-site to be served, it's essential to have a 'quarantine' area.

Breeders who visit a large number of shows throughout the spring and summer months will almost certainly have isolation facilities. Some of the shows are so close together that the 'standstill' rule would prevent them from going from one to the next if they did not have a self-contained building or a dedicated area in which to isolate.

Standstill rules do not apply if, for example, the pigs were housed in the isolation unit, moved off-site (eg to a show), and then returned to the same unit. This means that there are no restrictions on moving any other livestock.

Why isolate?
- It allows for the passage of the incubation period – the time it takes for symptoms of an illness to develop.
- It makes it easier to monitor, closely inspect or treat the pigs.
- It means more freedom for moving stock on and off the farm.
- It just makes good sense – why take risks?

Below: Biosecurity is paramount.

Above: Pigs in isolation.

Getting an approved unit
It doesn't cost anything to have a building or outdoor area approved as an isolation unit. All you have to do is contact your local Animal Health office and an inspection will normally be carried out by your own or another local vet, who will provide a written report that certain criteria can be satisfied:

- The unit must only ever be used for isolation.
- It should ideally be a building separate from any other housing livestock.
- The floors and walls must be in good condition so that they can be washed and disinfected.
- Disinfectant footbaths must be provided at the entrance.
- Staff must use dedicated protective clothing, and protective clothing should be provided for visitors.
- No other animals should come into contact with manure or effluent from the building.

OUTDOOR ISOLATION UNITS ARE POSSIBLE, BUT:

- Paddocks must only ever be used for isolation.
- They must be physically separate from any land or buildings used for other livestock.
- There must be a minimum distance of 3m (10ft) between the perimeter of the isolation paddocks and any other livestock.

ISOLATION UNIT

The pigs in this barn have recently arrived and are being isolated for biosecurity reasons – to ensure they are healthy before joining the rest of the herd.

Please disinfect your boots before entering the building and when you leave, and remember to wash your hands.

Potential health problems

You can scare yourself silly reading about all the various illnesses and ailments that pigs can get. However, if you're buying in weaners to fatten for meat, and you buy from a reliable source and care for them as you should, there shouldn't be too much to worry about during the few months they're with you. Having said that, it pays, of course, to be aware of what problems you might encounter – particularly if you see yourself becoming a long-term pig keeper in future, or if you plan to start breeding.

As mentioned earlier, if in doubt, always call your vet.

External parasites

Mange and lice can be found on even the healthiest pigs. Signs are repeated rubbing and scratching against any fixed object – sides of pens, arks, fence posts – often leading to loss of hair and broken skin.

Sarcoptic mange is caused by *Sarcoptes scabiei var suis*, a burrowing mite which works its way under the skin to lay its eggs. The skin reddens around areas like the eyes, ears and legs. It wrinkles and scabs begin to form, eventually leading to crusty patches. This results in severe irritation and can affect growth rates. The mites and eggs can survive for more than a month outside the body if conditions are right.

Typical signs of mange are rubbing and scratching and shaking of the head. Sows and gilts can go off their food, lose condition, and either be difficult to get in pig or else produce smaller litters; boars made uncomfortable by irritation may be reluctant to mate.

The only authorised treatments are ivermectin or doramectin. These can be given by vaccination. Alternatively, ivermectin can be given orally in the feed. The key to controlling the problem is breaking the cycle of infestation, so preventing transfer from one pig to another. Affected sows should be treated two to seven days before being moved into clean farrowing quarters. Once weaned, piglets should be moved into a mite-free area to prevent reinfestation.

Pig lice (*Haematopinus suis*) are blood-sucking creatures which are fairly big and can be seen crawling on the skin. They're more of a problem in small-scale herds nowadays than in large, commercial units. They're typically found on the hair behind the ears and between the legs – areas where

Below: Pig lice eggs.

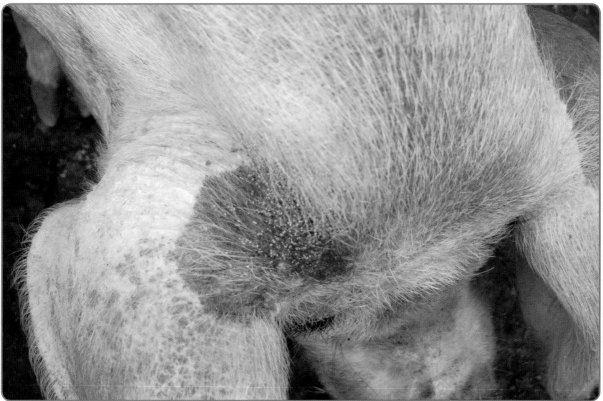

Alison Wilson

they like to lay their small white eggs. The eggs look like dandruff at first glance and are easy to miss.

Lice suck blood and transmit infections, and pigs can end up anaemic. In severe cases they can cause irritation and can have similar side-effects to mange – loss of condition and reduced fertility, etc. Unlike mange mites, lice cannot live more than a few days without a host, but they can be transferred to other animals in straw. The life cycle is between three and five weeks, all of which normally occurs on a pig.

There's a further problem with lice, in that they can easily spread disease – anything from swine fever to PRRS (see below).

Many pig keepers vaccinate as a precaution against both mange and lice, normally with a combined vaccine that also treats internal parasites. Alternatively, powdered or pour-on preparations may be prescribed by a veterinary surgeon, if considered appropriate.

Internal parasites

Intestinal worms can slow down growth rates and, in serious cases, can lead to death. Piglets should be wormed at weaning and the treatment repeated every six months if they're to be kept for breeding. This can be done by injection (as mentioned above) or by mixing pellets in with the feed. The drawback of mixing wormer with feed is that you cannot be certain that each pig has had the correct

Below: Pigs must be able to cool down in hot weather.

dose. The best time to inject in-pig free-range sows and gilts is a week before they farrow, thus breaking the cycle of infection to the piglet. Other pigs in the herd can be treated with pellets mixed in with their feed.

Regular worming and good management – including moving pigs to new pasture and allowing ground to rest – can help minimise problems. Worms can live for a long time in soil if the conditions are right, and pigs can become reinfected as they root and forage.

Signs of worms in young pigs include failure to put on weight, loss of weight, pot bellies, diarrhoea, coughing and dull-looking coats. Older, well-nourished pigs may have no obvious signs of worms, and a problem may not be apparent until slaughter time. Abattoir inspectors will condemn any livers that show 'milk spot' – lesions caused by larvae.

Your vet can take faecal samples from pigs of different ages, which can determine what kind of worm problem exists, how severe it is, and how it should be treated.

Heat-related problems

Heat stress can be a big problem when temperatures soar, because pigs aren't blessed with the sweat glands other livestock have, and therefore cannot cool down as easily. The pig's main way of losing body heat is through panting or immersing itself in water or mud. Food intake is reduced, so growing pigs fail to put on weight, while

Above: White-skinned pigs run the risk of sunburn.

lactating sows produce less milk. Heat stress symptoms include rapid breathing or gasping and the pig may tremble and have difficulty standing. Unconsciousness can follow and if not treated quickly enough the pig will die.

In extreme cases (eg if a pig's temperature is over 40°C) cold water from a hosepipe can save a life. Another approach is to soak a towel in cold water and lay it over the pig. Gentle sponging with cold water, particularly behind the ears, can help. Vinegar is often used behind the ears because it cools as it evaporates. Dehydrated pigs that cannot be persuaded to drink can be helped by placing ice cubes in the anus. Alternatively, cold water can be introduced into the anus using a 60–100ml syringe with a tapered nozzle, or, if this isn't available, a piece of rubber pipe and a jug of water.

When planning your pig paddock, make sure your pigs have access to a shaded area where they can avoid the sun, and if possible scrape out a mud wallow or leave a hosepipe trickling in one corner so that they can coat themselves in mud.

Light-skinned pigs can get sunburned very quickly. Pigs use mud like we use sun block, but areas with thinner hair – including the ears and the spine – can often be left exposed. In warm weather use a high-factor sun cream (50 SPF or higher) on vulnerable parts before they get burned. If the damage is already done, apply a soothing cream, but remember to reapply sun block before the sun comes out again, otherwise more damage could be caused. Badly burned pigs should be taken indoors for treatment and only returned to the field when fully recovered.

Lameness

This can be caused by a wide variety of things, including injuries, foreign bodies, bacterial infections, or simply uncomfortable flooring (eg ridged concrete surfaces at markets or shows). When a pig is reluctant to stand, preferring to sit like a dog, it can be a sign of arthritis – an inflammation of the joints, usually caused by bacterial infection.

Mycoplasma arthritis is a common occurrence. It exists in most herds, but not all pigs will be affected. It's caused by a microorganism, *Mycoplasma hyosynoviae*, which lives in the upper respiratory tract and the tonsils. Some herds may have it without exhibiting any ill effects, while in others it can cause serious problems. Growing pigs aged 8 to 30 weeks are often affected – particularly gilts soon after they arrive at a new farm, or early in pregnancy. Older sows are unlikely to display symptoms because they'll have developed immunity from repeated exposure.

A typical symptom is stiffness in the rear legs. Pigs are reluctant to get up, but manage to shuffle forward to reach food or water. Treatment is by a series of injections with a specific antibiotic proven to be effective (eg lincomycin or tiamulin). Normally four injections, on consecutive days, are recommended. If *Mycoplasma hyosynoviae* is the cause the pig should show signs of recovery within 24 to 36 hours. Treatment is most effective if given early, and can be given as an in-feed medication. Painkilling injections will also greatly improve the pig's response to treatment.

Below: Sow and litter enjoying a mud wallow.

Below: 'Dog-sitting': a symptom of Mycoplasma arthritis.

Mastitis

Lactating sows can suffer acute mastitis – inflammation of the mammary glands – as a result of bacterial infection, either through the opening in the teat or as a result of an abrasion, possibly caused by a piglet's sharp teeth. Good hygiene in farrowing areas is essential to minimise risk.

Signs of acute mastitis can appear within 24 hours of farrowing. The udder becomes red and swollen, hard or lumpy, and hot to the touch. The sow may lose appetite and have a raised temperature, and, because of her discomfort, be lying on her udder, preventing the piglets from suckling. Denied feeds, the litter will gradually become dehydrated and weak and will starve. Delayed treatment can result in the sow's death.

Treatment is by injections of appropriate antibiotics and non-steroidal anti-inflammatory drugs. Oxytocin is sometimes given to help drain milk and remove bacterial toxins from the udder, and the milk will have to be stripped out by hand.

Chronic mastitis is common and can affect one or more glands. It's usually seen after weaning, when the udder stops producing milk and shrinks back. The glands can be hard, swollen and lumpy, and ulcers may occur. Although lesions may heal, problems can reoccur and worsen at the next farrowing. Udders can become so swollen that they drag on the ground and can be trodden on. Antibiotics can be given, but severely affected sows should be culled.

Below: Sows may lie on their bellies to avoid feeding piglets.

Bob Stevenson

Above: Mastitis is a painful condition.

Poisoning

Salt poisoning – or, to give it its proper name, water deprivation – can occur quickly and can prove fatal within a short time, whatever the age of the pig. Water is vitally important to the pig's diet. Adults drink vast quantities, with nursing sows drinking around 20 litres a day – more in extremely hot temperatures. As a rough guide, a pig needs to drink more than twice the amount of water as the weight of food it eats, eg a pig eating 2kg per day will need to drink at least 4.5 litres of water per day.

A shortage of water causes the normal levels of salt in the diet (0.4 to 0.5%) to become toxic. Pigs initially lose their appetites, so always check the water supply if a pig isn't eating. As the condition develops, fits and lack of coordination may be seen, with pigs wandering around as if blind. They may also press their heads against walls or other solid objects.

Rehydrating the pig – by trickling small amounts of water into the mouth or into the rectum – is the first step, though treatment is not always effective.

Pigs will have a go at eating practically anything, so keep potentially dangerous items – such as discarded rubbish or anything believed to be toxic – well out of their reach. Plants like foxgloves, hemlock, laburnum and yew are well known as being dangerous to livestock, so check the paddock regularly. Watch out for greedy pigs overdoing it on fallen acorns, which can cause stomach problems and can even make pregnant sows abort. Another thing to be wary of is allowing your pigs to forage unchecked in an orchard. Windfalls eaten by pigs can start fermenting and cause drunkenness and stomach upsets.

PMWS and PDNS

Post-weaning multi-systemic wasting syndrome (PMWS) and Porcine dermatitis and nephropathy syndrome (PDNS) are diseases that are linked to the same infectious agent, Porcine Circovirus Type 2 (PCV2), a hardy virus which is resistant to heat and also to most disinfectants. Signs of both PMWS and PDNS can be confused with other diseases, including classical swine fever, so it's important to contact your vet immediately.

PMWS can strike any herd, regardless of its size, health status and whether indoor or outdoor. It generally affects pigs between 6 and 14 weeks old, causing lack of appetite, depression and weight loss. Pigs appear pale or jaundiced and have diarrhoea, respiratory problems and sometimes mild conjunctivitis. In some previously healthy-looking pigs, sudden death occurs.

The most recognisable clinical signs of PDNS are skin lesions, which are red/brown. These appear on the ears, face, flanks, legs and hams, normally in pigs from 10 to 16 weeks old, but sometimes as early as 5 weeks and as late as 24 weeks. Pigs lose condition and die.

Despite the availability of effective preventative vaccines, both diseases are still widespread. Around 80% of pigs affected die or are destroyed for welfare reasons. However, some treatments are recommended:

- injecting with long-acting antibiotics at the first sign of disease;
- adding a broad-spectrum antibiotic to drinking water to control secondary infections;
- injecting pigs with corticosteroids and antibiotics to reduce lung infections and congestion.

Experts agree that the best way to keep these two diseases under control are to vaccinate; to limit contact between groups of pigs; to keep stress to a minimum; to keep buildings and equipment scrupulously clean; and to make sure that pigs are well nourished to help their immune systems.

Bob Stevenson

Above: A piglet suffering from PMWS.

Right: A pig with classic skin lesions caused by PDNS.

Bob Stevenson

PRRS

Porcine reproductive and respiratory syndrome (PRRS) was first recognised in the USA as 'mystery swine disease' or 'blue ear disease'. The virus is spread by nasal secretions, saliva, faeces and urine, and can be airborne for up to 3km (two miles).

Signs of PRRS vary considerably from herd to herd. First to be affected are dry sows, lactating sows and suckling piglets. A sow may have short periods when she goes off her food, or eats only a small amount. Her temperature may be raised and she may lose a pregnancy close to her farrowing date. Some sows farrow slightly early or return to service 21 to 35 days after mating; in others, oestrus may be delayed. Sows that farrow might, in addition to losing appetite, be reluctant to drink. They may suffer from lack of milk and mastitis, and can have increased stillbirths of large mummified piglets; live offspring can be extremely weak and unable to reach a teat to suckle. A sow's ears may become discoloured – hence the name 'blue ear disease'. Coughing and other respiratory signs may be seen, and even pneumonia.

Piglets are usually only affected when the virus enters the herd for the first time. Litters can be affected for several weeks before sows develop an immunity and pass on protection via colostrum. Newborns are likely to be affected by pneumonia, coughing and scouring, together with general weakness. The disease can also affect piglets soon after weaning, when the natural protection passed on by the sow starts to wane. Coughing and pneumonia are two typical symptoms, along with lack of appetite and gradual wasting.

Signs in adult boars include loss of appetite, increased temperature, lethargy and loss of libido. Infection can reduce fertility, resulting in small litter sizes.

There's a PRRS vaccine available for sows and gilts that reduces reproductive problems. In addition, it's important to control secondary bacterial infection, so antibiotics will also be given.

Enzootic pneumonia (EP)

This disease is caused by the microorganism *Mycoplasma hyopneumoniae* and attacks the lower areas of each lung lobe. Common in pig herds of all sizes, it can be serious, especially when other infections are present. It can be transmitted by the movement of pigs, but also can be carried by the wind for up to 3km (two miles).

The acute form of the disease is normally only seen when it gets into a herd for the first time. Symptoms include coughing, breathing difficulties, fever and death. When the disease has been present for a long time sows develop antibodies that pass to their piglets and protect them.

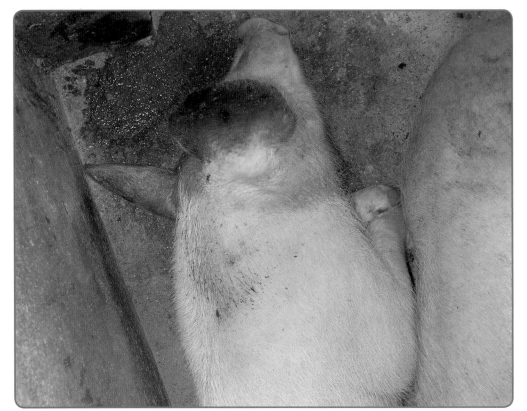

Left: PRRS causes discolouration of the ears – hence the name 'blue ear disease'.

Bob Stevenson

Right: Poor husbandry allows infections to flourish.

However, when the piglets are two to four months old the protection wears off, and coughing begins. Vets describe the coughing as 'non-productive' – a dry cough, repeated, which doesn't bring up any mucus. Growing pigs may lose condition and slaughtered pigs with EP will have lesions on the lungs.

Treatment is by antibiotics, and there are also several vaccines available, so discuss a treatment programme with your vet.

Better husbandry can help: if you have EP-free pigs, any new ones brought in should not be mixed with your existing herd until they've been in isolation for the incubation period (two to eight weeks); buildings should be well ventilated and not overcrowded.

Below: Scouring.

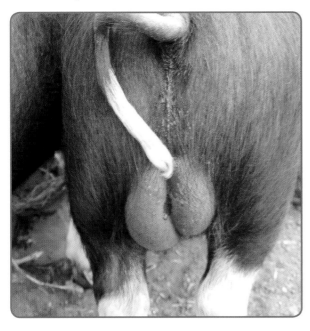

Scouring

Piglets are often affected by scouring (diarrhoea). It can be caused by a whole range of things, from bacterial infections to a change in diet or overfeeding. Bacterial scours can kill piglets very quickly indeed. When piglets are born, the intestinal tract has very little immunity to disease-producing organisms. Colostrum from the mother provides some protection, but bacterial infection, coupled with poor environmental conditions, can overcome immunity.

Good husbandry and hygienic practices (replacing bedding regularly, thoroughly disinfecting pens and equipment) are key factors in keeping problems at bay. Draughty conditions can exacerbate problems.

Piglets can also experience scours at weaning, as the stress of being separated from the mother and the switch from milk and solid food to solid food only, can make them more susceptible to viral and bacterial infections.

Antibiotics can be effective, and mixing an electrolyte solution (sodium, chlorides, potassium, calcium and bicarbonate) in water can help replace essential salts and prevent dehydration. There are also vaccines that protect against scouring.

Swine influenza

Caused by a number of closely related influenza viruses that can modify to produce new strains. Infection can cause infertility, death of embryos, and small litter sizes. Fertility of boars can also be affected. Symptoms include high temperatures, coughing and pneumonia. A particular H1N1 influenza virus, known to be highly pathogenic, caused a pandemic affecting many countries in 2009. It was inaccurately dubbed 'swine flu' by the media. However, pigs were not involved in spreading the disease to humans.

Prevention

If you bought your pigs from a good breeder, they should have been wormed at weaning and their mothers should have been vaccinated against some of the main diseases, providing them with a degree of immunity for the first few months. If you're keeping them longer, there are a number of things which you should consider taking pre-emptive action against.

Erysipelas

If you only vaccinate against one thing, make it this one. Erysipelas can cause everything from sterility and reproductive problems to sudden death – but it's preventable, and more and more breeders are now vaccinating against it.

The bacterium that causes erysipelas, *Erysipelothrix rhusiopathiae (insidiosa)*, is carried in the tonsils of almost all pigs and is transmitted via saliva or faeces. It can also live in the soil for several years, and can be carried by birds and other animals. The disease can be caused simply by the bacterium, but viral infections can also trigger it. Bacteria may enter the bloodstream in a number of ways, eg through the wall of the digestive tract or a skin abrasion. Septicaemia (blood poisoning) then develops within 24 to 48 hours.

In the most serious cases pigs are found dead, or running high temperatures. The telltale skin lesions

Below: Classic Erysipelas rhomboids.

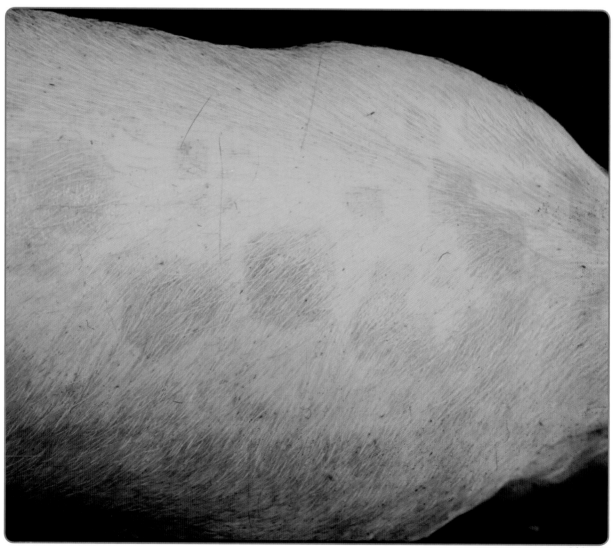

Bob Stevenson

Safety when dealing with pigs

If a pig is showing clinical signs of disease or has been in contact with diseased pigs, handlers should follow these precautions:

- Ensure cuts and abrasions are covered with waterproof dressings.
- Wear gloves and overalls at all times.
- Remember to disinfect protective clothing after use.
- Wash your hands after handling the affected animal or by-products, even if gloves were worn.

In addition, anyone involved with handling pigs or working in the same environment should be vaccinated against tetanus.

associated with the disease are small, red rhomboids or diamond-shaped lumps measuring 10mm to 50mm (½in to 2in) all over the body, which may turn black as the disease progresses.

In less acute cases skin lesions may appear, pigs have a high temperature (40–42°C, 104–108°F), and they may be lethargic and have a lack of appetite, but they won't necessarily appear ill. Pigs may be thirstier than normal and be breathing faster than normal. The joints may be affected, causing chronic arthritis and lameness.

If caught early enough penicillin can bring a rapid response, but often the symptoms are not recognised in time. A preventative vaccine should be given to both gilts and boars every six months. Many vets will recommend giving a combined erysipelas and porcine parvovirus (PPV – see below) vaccine to young breeding stock, followed by six-monthly doses of the normal erysipelas vaccine.

Below: Penicillin can help if signs of erysipelas are spotted early.

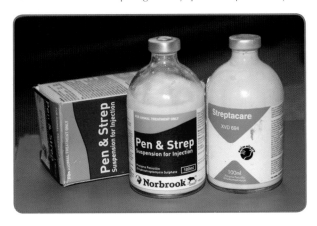

Porcine parvovirus (PPV)

This is another important one to vaccinate against if you're breeding pigs and are likely to have your herd come into contact with pigs from elsewhere, eg bringing in or hiring out boars; buying at markets; exhibiting at shows, etc.

This is the most frequent cause of infertility and farrowing problems, resulting in phantom pregnancies, small litters, low-birth-weight piglets, stillborns and – the most recognisable sign – mummified piglets of varying sizes. Resistant to disinfectants, this virus multiplies in the intestinal tract without causing any obvious symptoms. Like erysipelas, it can live in the environment for a long while. The combined erysipelas and parvo vaccine offers life immunity from parvovirus – but not from erysipelas (see above).

Below: Mummified piglet caused by PPV.

Notifiable diseases

There are a few diseases which are considered such a threat that you have to notify your local Animal Health office immediately should you suspect that your animals have any signs of them. The ones relating to pigs are:

Anthrax
Normally associated with cattle and sheep, it can affect all animals quickly and can be passed on to humans. Symptoms include high temperature, fever, lack of appetite, breathing problems, swollen throat and sudden death.

Aujesky's disease (pseudo rabies)
Caused by a herpes virus, symptoms include coughing, sneezing and difficulty breathing, along with fever and weight loss. Gilts and sows can have reproductive problems like abortion, stillbirth and mummified foetuses. Piglets also show lack of coordination. The last outbreak in the UK was in 1989.

Below: Vehicles should be disinfected before moving from farm to farm.

Foot-and-mouth
Lameness, salivation, and blisters on the feet, snout and tongue are the main signs. There were a handful of cases in south-east England in 2007, but the last major outbreak was in 2001 and led to mass slaughter and movement restrictions across the UK.

Rabies
Rare in pigs, this is a fatal disease of the nervous system spread by saliva from the bite of an infected animal. Signs include nervous twitching, salivation, paralysis, fear of water and aggression.

Swine fever
Both African swine fever and classical swine fever are notifiable. They are caused by very similar viruses and can only be identified by lab tests. Symptoms in sows include high fever, stillbirths, malformations, convulsions and death. Piglets born can be weak and trembling, with poor coordination, high fever, vomiting and diarrhoea.

Alamy

Above: Diseases like foot-and-mouth can shut down the countryside.

Monitoring your pigs' health

You need to learn to 'spy' on your pigs! Obviously, pigs react differently when they're left alone in their group and when they see you and realise it's feeding time. So try to observe them without them realising you're there, when they're relaxed.

The first thing to work out is what is normal behaviour for your pigs. Unless you can recognise 'normal' you won't be able to recognise 'abnormal'. These are the things to look for in healthy pigs:

- Are they eating and drinking normally?
- Are they alert and vocal, and are their eyes bright and clear?
- Do they look well nourished, but not overweight?
- Are their faeces consistent with what you're feeding them?
- Do they have straw-coloured urine?
- Are they reactive and moving around freely?
- Is their body temperature within the correct range of 38.6°C to 38.8C° (101.5°F to 102°F)?
- Is their breathing rate normal (20 to 30 breaths per minute in adults, 50 in piglets)?
- Is their skin clean and free of lesions?
- Are their membranes (eg gums, vulva) salmon pink as they should be? Paler than normal or blue-tinged could suggest anaemia, while yellow could indicate a liver disorder.
- Is there any unusual discharge from the nose or vulva?

Swine vesicular disease

Signs are almost identical to foot-and-mouth and tests have to be carried out to determine which is which. The UK has not seen a case since 1982.

Teschen disease

A nervous disease that causes fever, loss of appetite, lethargy and loss of coordination. Pigs may grind their teeth, smack their lips and squeal as if in pain. Paralysis occurs, followed by death within three or four days. This disease has never been recorded in the UK.

Vesicular stomatitis

Vesicular diseases in pigs can be caused by infection with foot-and-mouth disease, and vesicular stomatitis has identical symptoms. Laboratory diagnosis is the only way to determine which disease it is.

Zoonotic infections

These are diseases that may be present in animals or animal by-products and can be transmitted to humans, ranging from mild food poisoning to some of the notifiable diseases. Unlike other species, pigs have relatively few serious zoonotic organisms, but anyone coming into close contact with pigs or areas where they've been should be aware that general safety precautions should be followed to avoid contamination.

Administering injections

If you're going to keep pigs for breeding you'll need to learn how to give medication by injection. There are three techniques: subcutaneous (under the skin), intramuscular (into the muscle), and intravenous (normally into the veins of the ear, but also into the jugular vein and the large vein that leaves the heart). Which technique is required depends on the product being used, but beginners need only concentrate on the first two, as intravenous injections are best left to the vet.

Ask your vet to show you how to give vaccinations and other routine treatments and make sure you're sufficiently confident before attempting a procedure unsupervised:

- Always read the instructions on a veterinary product before administering and ensure that you know which type of injection technique is required.
- Ensure that injection equipment is clean and sterile before use.
- Always use the correct size needle for the job (see table)
- If you're injecting a number of pigs with the same product, change needles after every five to ten pigs.
- To reduce the risk of contamination, always keep a sterile needle in the bottle to refill the syringe, then use a different needle to inject the pigs.
- The animal being injected must always be adequately and safely restrained.
- Note the withdrawal period – the time between the last dose of medicine administered and the time when it's safe for the animal to enter the food chain.
- Record everything in your medical records book.

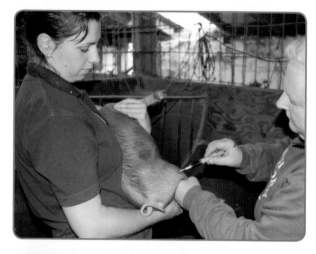

Above: Weaner restrained for injection.

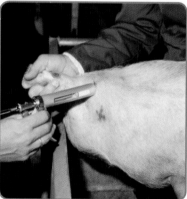

Bob Stevenson

Left: How to give an intramuscular injection. The needle is in the correct position; the red cross shows a spot commonly used, which is too low.

Sites of Injections

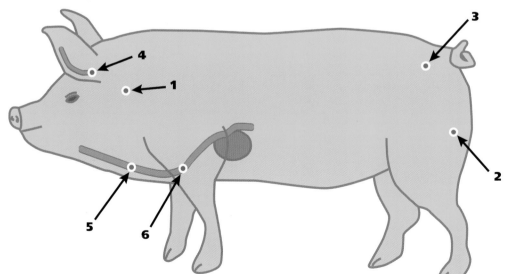

1. Site for subcutaneous or intramuscular injections
2,3. Sites for intra muscular injection (piglets only)
4. Site for ear vein for intravenous injection
5. Jugular vein
6. Site of anterior vena cava

Subcutaneous

There are three sites for subcutaneous injections in pigs, depending on size. In small pigs the best place is under the fold of skin on the inside of the thigh, or under the skin behind the shoulder. In growing or adult pigs, the best site is approximately 25–75mm behind and level with the base of the ear. The skin should be pinched between finger and thumb and the needle inserted at a 45° angle. Care should be taken to ensure the needle doesn't go right through the skin and out the other side.

Intramuscular

The preferred site for weaners, growers, finishers and adults is 50–75mm behind the ear. If the injection is given too far back the drug could be deposited in fat instead of muscle; if too low, it could end up in the salivary gland. Small piglets are often injected into the ham of the hind leg because there's not much muscle on the neck. This is not recommended for larger pigs because of the possibility of abscesses, and because it causes bruising to a valuable part of the carcass.

Which needle do I use?

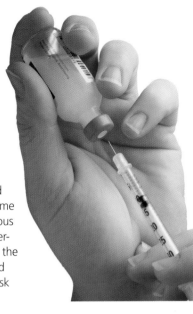

This table provides a guide to the kind of needles required for intramuscular injections. For subcutaneous injections, a shorter needle is required – between 12mm and 25mm (0.5–1in). Some drugs are more viscous and will need a wider-gauge needle. Read the instructions enclosed with the drug and ask your vet for advice.

Pig weight	Needle length	Gauge/thickness
Up to 10kg	12–18mm (0.5–0.75in)	20 to 21
10 to 30kg	18–25mm (0.75–1in)	18 to 19
30 to 100kg	25–30mm (1–1.25in)	18
100kg and over	38–44mm (1.5–1.75in)	16

Essential medical kit

- A snare for restraining pigs.
- Antibiotics.
- Needles and syringes of varying sizes.
- 'Sharps' box for disposing of needles.
- Thermometer.
- Erysipelas and parvo vaccine.
- Worming products – injectable or pelleted.
- Antiseptic cream.
- Wound powder.
- Injectable iron.
- Disinfectant.
- Cotton wool.
- Surgical spirit.
- Iodine.
- Scissors.
- Electrolytes solution.

Right: A snare or 'twitch'.

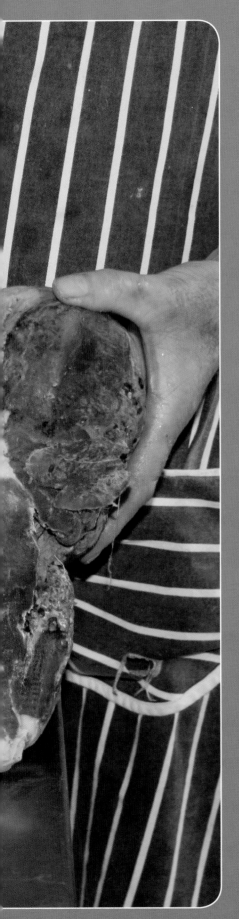

EATING
YOUR OWN

Eating your own

Few things can beat the satisfaction of producing your own food, and eating your own home-raised meat is a particularly fulfilling experience. Once you've tasted free-range pork that has been produced with love and care you'll never go back to the cheap, fast-grown, flavourless stuff that you find in most supermarkets.

But when the time comes, will you be ready for D-Day ('D' standing for 'death', of course)? Many people start off with the full intention of raising livestock for the table, but grow so attached to their animals that they can't let them go when the time comes.

It's a sad fact that the majority of people have grown up never making a connection between the animals they see in the fields and the sausages that end up on their plates. Some will only buy meat that appears in neat packages on the supermarket shelves and bears no resemblance to the animal it came from. Rearing your own livestock completely removes that detachment. You're responsible for that animal's care and welfare; the power of life or death is in your hands.

So are you psychologically prepared to eat something you've fed, watered and cared for, day in, day out? It may not just be your feelings that you need to consider. How will the others in your family feel about sending the pigs off for slaughter? Children may get fond of them and give them names, so you need to make it clear from the start that these particular pigs aren't pets, and won't be around forever.

So if you're contemplating raising meat for the table, you should think long and hard before making the decision. If you don't, you could end up with some very expensive pets that could live to as long as 15 or 20 years.

Below: Roast leg joint.

Planning ahead

Abattoirs aren't as widespread as they once were. Just a few decades ago there used to be at least one in every town or village, but these days finding one on your doorstep isn't very likely. You'll almost certainly need to book in advance. Turning up on the day isn't possible either. Abattoirs aren't like shops, which open up every day in the hope that someone will turn up and give them some business. Some abattoirs reserve different days for different species, and almost all will have a booking system.

So the message is, don't wait until your pigs are at slaughter weight and then start wondering what you're going to do with them. Do your homework well in advance and make sure you understand what the abattoir will accept. Some will only take pigs within a certain size range, because their facilities aren't designed for either underweight or supersized pigs – so find out before it's too late.

When you contact the abattoir, you'll be asked whether you just want your pigs slaughtered, or want them butchered as well. The latter is known in the trade as 'kill and cut'. The standard of butchery at abattoirs can vary, so ask others who have used the service for their opinion.

The other straightforward option is to get the pigs killed and delivered to a butcher, or to cut them up yourself. It's assumed, at this stage, that the meat you're producing is for your own consumption and not for sale to others. If you're intending to sell your pork, the business of who cuts your meat gets more complicated. Premises where meat is butchered for sale to a third party have to be licensed and supervised by the Meat Hygiene Service. However, local authorities have discretion to allow cutting at unlicensed premises under certain circumstances, but check with the Food Standards Agency for more information, because this is a very grey area.

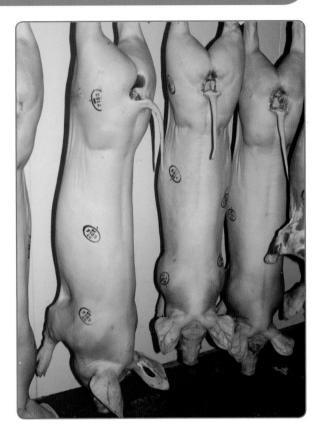

Above: Carcasses hanging before being butchered.

Below: The bottom half of a carcass about to be cut into leg joints.

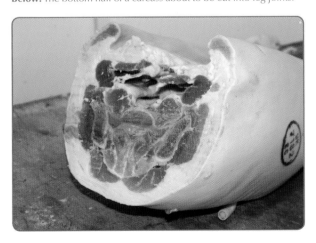

Humane slaughter

It's important that when you start keeping pigs, you think about the end of the process too. Animals have to be slaughtered humanely and, if they're for anyone other than yourself, they need to be killed at a licensed abattoir. The Food Standards Agency has a list of licensed premises, and produces a range of publications with advice on food production. Go to their home page at www.food.gov.uk/ and look for the link, 'Food industries'.

Home killing is an option only if the meat is solely for your own consumption – but it certainly isn't something for beginners. The Humane Slaughter Association is an excellent source of information. It produces a range of publications, all available by mail order. There are also DVDs and videos – though not for the squeamish – showing exactly how an animal should be killed.

From piglet to plate

If you buy your pigs at eight weeks old you'll only have them for a few months. Traditional breeds reach pork weight at between five and six months old, and will be ready for bacon at between eight and ten months old. Modern or cross-bred pigs grow even faster, so you could have your meat even earlier.

Judging when pigs are ready for slaughter can't always be done on age alone. The breeder you buy from may kill pork pigs at six months old, but growth and development can be affected by numerous things. You may be feeding different amounts, or using feeds with different protein contents. Indoor pigs will fatten much quicker than ones that are free-range and using up their energy outdoors. Temperature is a factor too: in very cold weather, a pig uses much of its food to keep warm; in hot weather, it may go off its food and lose condition.

Generally speaking, pigs fall into three carcass weight categories:

- **Porker** – A pig aged anything from four to six months (depending on the breed) and weighing around 55kg (120lb) when ready for butchering (which is also known as 'dead weight').
- **Baconer** – A pig reared until eight or ten months old, and weighing about 80kg (175kg) dead weight.
- **Cutter** – A pig somewhere between pork and bacon weight, kept until about nine months old to produce larger-size pork joints.

With experience you'll be able to tell when your pigs are ready just by looking at them. You'll come to know how big you like your joints and chops and how much fat you want to see on your meat. You'll soon be able to make adjustments accordingly and alter your feeding regime or slaughter dates to produce the results you require.

In the meantime, however, you may need a little help. Purpose-made weighing crates are expensive, but there are other ways of estimating a pig's weight. The simplest is by using a 'weigh band' – a tape measure that converts the 'chest' measurement of your pig into the corresponding dead weight. Just pass the tape under the 'armpits' of your pig (ie just behind the front legs) and over the shoulders, and see what the tape says. It helps if the pig is distracted by eating when you do this. The tape will give you a useful guide, but bear in mind that if you have a breed which doesn't have much of a back end (eg a Tamworth), the actual dead weight will be less, so subtract about 6kg to compensate.

A weigh band is a fairly cheap investment, but you could also try this traditional method:

1 Measure around the pig as before.
2 Multiply the measurement by itself.

Left: Using a weigh band.

Left: A pig weighing crate

3 Measure the length of your pig – from the top of the head, between the ears and down to the point where the tail sprouts from the body (not to the end of the tail).
4 Take the result of step two and multiply by the length of the pig.
5 Multiply by 69.3 for a weight in kilos if you've taken your measurements in metres, or divide by 400 for a weight in pounds if you've used inches.

For example, if the pig's 'chest' measurement is 1.2m and its length is 1.01m, step 2 would give you a figure of 1.44; multiplied by its length of 1.01 (step 4) would give you 1.4544; and this multiplied in turn by 69.3 (step 5) would give the pig's dead weight as 101kg.

Preparing for the abattoir

Withdrawal period

One thing you must consider before booking a slaughter date is whether your pigs have received any medication. Veterinary drugs have different withdrawal dates (ie the amount of time you have to allow before slaughter to make sure the medication is completely out of the animal's system), so check your medical records book to make sure.

Identification for slaughter

Each pig will need to be identified with a tag or a 'slap mark' bearing your herd number. Metal tags are recommended, because not all plastic ones will withstand the temperatures used in the abattoir process. If you haven't ordered your tags or slap marker, do so at least a couple of weeks before the pigs are due to go in order to allow for delivery hold-ups. No abattoir will accept your pigs unless they're properly identified. You might find it easier to tag them once they're loaded in the trailer. Ear tags inserted too early often cause problems, as young ears continue to grow and the tags will become tight and uncomfortable, and sometimes infected. Slap marks are best applied close to the slaughter time; too early and they may be obscured by hair growth. There should be one slap mark on each shoulder.

Paperwork

The eAML2 needs to be set up in advance of any movement of pigs. As mentioned in Chapter 3, there is a drop-down menu which allows you to choose the type of movement required. Print out a copy of the form and take it with you to the abattoir.

Loading up

Spending time with your pigs makes them easier to handle. If you train them to respond to the sound of a food bucket being rattled, it should help when it comes to loading them into the trailer. It also helps if you get them used to the trailer beforehand, maybe by taking it into the pigs' field a few days before you need to move them. Let them explore it in their own time and feed them inside it and this should help remove the fear of climbing up the ramp.

If you have to take the pigs to the trailer, rather than the other way round, don't feed them much the night before. Then, by the time it comes to load, they should be hungry and ready to follow a bucket.

You should only give your pigs enough food to entice them in. Abattoirs don't like dealing with animals that have full bellies.

Get your trailer into an area which can be securely fenced off using sheep hurdles or other barriers and create a kind of corral – a holding pen from which there's no going back.

Tempt your pigs out into the holding pen and make sure you shut them in securely. It helps to have a few helpers – one to rattle the food, one to bring up the rear with a large pig board or sheep hurdle, and another ready to get the tailgate of the trailer closed as soon as the pigs go up the ramp. It's much better to let the pigs walk up in their own time, rather than trying to force them on board. Just make sure the holding area is secure – and try to be patient!

Pigs often dislike the way the trailer ramp feels beneath their feet, so try covering it with straw or a piece of old carpet.

Checklist before setting off

- Are the pigs fit and well enough to travel?
- Do they have metal tags in their ears/are they slap-marked?
- Have you completed all the relevant sections on the movement licence?
- Is your trailer clean and well ventilated, and is the tailgate securely closed?
- Are all the electrics on the trailer working?
- If you're travelling more than 65km (40 miles) or the journey is going to last more than eight hours, take your certificate of competence with you.

Below: Guide your pigs using a board and stick.

Arriving at the abattoir

If you lack confidence with a trailer, it might be worth taking a drive to the abattoir beforehand so that you can do a recce and see where you'll be offloading. At some abattoirs you can drive straight into the lairage (the offloading and holding area), while at others you have to reverse in. If you think you'll get stuck, don't panic – there will usually be someone willing to do it for you.

When you arrive you'll normally be met by a member of staff from the abattoir – possibly the slaughter man himself – and a vet from the Meat Hygiene Service. The vet has a responsibility to inspect the animal before and after it's slaughtered, and to ensure that any meat entering the food chain is fit for human consumption.

Below: A vet inspects every arrival.

Guilt pangs

If you didn't have a slight crisis of conscience at this point, you wouldn't be human. You'll have been feeding and spending time with your pigs for several months and building up a relationship with them, so it's natural to feel a little guilty. It may not be easy but, when you open the trailer try not to get too sentimental. Remember why you got pigs in the first place – to raise great-tasting pork in a welfare-friendly way and to have some degree of control over what goes into your food. Those pigs have enjoyed a really good life – far better than a poor creature that's

Below: Slaughterman takes the pig to a holding pen.

Above: Waiting for slaughter in a holding pen.

been reared indoors, and been denied the light of day and the freedom to root about.

If you really feel you can't cope with the last goodbye, leave the job of getting your pigs out of the trailer to the abattoir staff. They'll have seen it all before and won't mind. Just keep remembering that in a few days' time you'll be feeling a whole lot better. You'll be collecting your very own meat and looking forward to that first fantastic roast.

You must clean your trailer within 24 hours of leaving the abattoir, or before it's next used for carrying animals, whichever is the sooner. Larger abattoirs will offer washing-out facilities, but smaller ones might not. If you choose to clean it at home instead you'll be asked to sign a form giving an undertaking to wash and disinfect it. You'll receive a copy and the abattoir will keep one.

Abattoir procedure

Your pigs will be put into a holding pen – the lairage – until staff are ready to slaughter them. Each pig is taken through to an area where it's stunned with electrified tongs placed either side of the head. This causes an epileptic fit that 'short-circuits' the brain and renders the animal unconscious. It's shackled by one hind leg and hoisted off the ground and then the major blood vessels are cut, causing rapid blood loss and bringing a swift death.

The carcass is then transferred into the hot water vat, which loosens the hair before the skin is scraped and brushed clean. At the next stage of the process, the carcass is split open for the stomach and the 'pluck' (the tongue, oesophagus, heart, lungs, liver and diaphragm) to be removed and separated and inspected by the vet. Occasionally certain organs will be considered unfit for consumption, eg the liver sometimes has 'milk spots' – small lesions caused by worm larvae. After being inspected, the carcass is chilled and hung before being butchered.

Briefing the butcher

Wherever you resolve to have your meat cut, you have to decide what you want done. If you leave it to the cutter's discretion you could end up with 'Christmas joints' – big enough to feed a family of ten – when what you might need are small pieces, suitable for two to four people.

Unless you specifically ask for the head, internal organs and the trotters, some abattoirs (or whoever is doing the cutting for you) will assume you don't want them. There's an old adage that you can use every part of a pig apart from the squeal, so if you want the liver, intestines, kidneys, heart, lungs and so on, tell the slaughterman that you want the 'pluck' back. But be warned – it will probably come back in one huge plastic bag and will take some sorting out!

Specify how you want everything packed. Some places put everything on polystyrene trays with plastic film on top. One drawback with this is that blood can

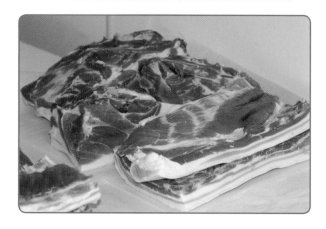

Above: Shoulder (left) and belly joints.

Below: This is what just half a pig will look like when butchered – so make sure you have room in your freezer.

Above: These boxes contain an entire butchered pig, minus the head.

ooze out of the meat and eventually out through the end of the film. Secondly, sharp bits of bone can easily break through the film, exposing the meat, and it can get 'freezer burn' – a drying-out process which leads to discolouration and affects quality. Vacuum-packing works best; it looks a lot nicer and keeps the meat protected for much longer.

You'll probably be asked if you want sausages. The meat from the cheeks is often used for this, along with some taken from the belly and shoulders. However, sausages won't normally come as part of the basic price, so make sure you know how much you'll be charged. The average rate has long been '£1 a pound' (250g), though prices will undoubtedly vary. It's worth asking for the sausages to be divided up into manageable-size bags, or in half-dozens on trays.

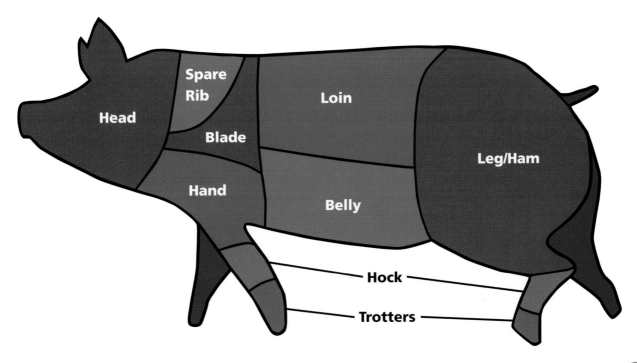

Making your own sausages

Sausages made from your own meat are better than any bought in shops, and you might fancy trying to make some yourself. Processing your meat in this way is a great way to add value to your pork, particularly the cheaper cuts. Lots of customers would never think of cooking a joint of pork apart from on a Sunday, but sausages will make an 'anytime' meal – whether for breakfast, lunch or dinner – so they always sell well if they taste good.

However, as mentioned earlier, if you're planning to sell to others there are health, hygiene and licensing regulations to be borne in mind, so contact the FSA and your local Trading Standards office for up-to-date guidance.

Making sausages is something which many people fancy trying their hand at, but relatively few get round to doing. It may be fiddly and frustrating to start with, but it can be immensely satisfying once you've got the hang of it. If you don't fancy sausage making, burgers are easier and every bit as tasty – just mince, add seasoning and cook.

Useful equipment

You don't need to invest in hugely expensive equipment when you're just starting off, but bigger, faster, more robust mincing machines will undoubtedly make the job a lot quicker and easier. The same goes for sausage-stuffing machines. However, there's nothing wrong with buying a small domestic mincer that has a sausage-making attachment and seeing how you get on.

You can put pretty much anything you want in a sausage, but you'll need a good proportion of fat. Most butchers work on the basis of the meat content being half lean, half fatty. Traditional rare breeds are perfect for sausages, being blessed (or cursed, if you talk to butchers) with a better covering of fat than the more modern and commercial breeds.

Traditionally sausage makers used breadcrumbs or rusk to bulk out the mixture and save meat, but if you're making your own you can choose what – if anything – you add, and how much. In defence of breadcrumbs, they do help hold fat within the sausage and therefore retain flavour, and they can also improve texture. For this reason you'll find that most ready-made sausage seasoning mixtures will include some kind of cereal. Of course, you may want to make your own sausages 100% meat – particularly if you have potential customers who are keen to source gluten-free products – but be prepared to experiment with different recipes to achieve the texture you require.

Sausage casings

Casings, or skins, can be natural 'hog casings' made from intestines, or man-made ones which contain collagen from cow hide as well as preservatives. There are pros and cons with both: the collagen ones can be used straight away, have a long shelf life (up to two years) and don't need to be refrigerated; the natural ones need to be soaked for a few hours – preferably overnight – before use, and must be rinsed before they're used. They aren't as long-lasting as the collagen casings, but keep for about two months if they're salted and refrigerated. Hog casings may have a more natural texture and give a better-looking finished produce when using traditional, coarser mixes or making really big bangers.

Left: Sausagemeat can be flavoured with practically anything that takes your fancy.

How to make sausages

Books can be very useful when you're learning a new skill, but there are some things you need to learn first-hand from a real person. Sausage making is one of them. The best thing to do is either to get yourself on a course, or persuade someone with the necessary know-how – maybe your local butcher – to show you how it's done. The basic steps are as follows:

1 Dice your meat into chunks and put it through a mincer. This mincer – being used at Food Centre Wales in Horeb, West Wales, by award-winning sausage maker Dennis Elliott – is a professional, heavy-duty machine, capable of dealing with large quantities of meat, all day every day. Small domestic mixers can be used just as effectively with more modest quantities of meat.

2 The mince is put in a bowl with seasoning and mixed thoroughly with clean, bare hands. It's then returned to the mincer a second time to achieve a finer consistency.

4 The resultant long sausage is pinched and twisted at regular intervals to create individual sausages, which are then 'linked'. This is the really tricky bit, and needs to be observed rather than explained. If you can't get on to a course, look for a sausage-making DVD or search the Internet for demonstrations (www.youtube.com has several examples).

3 The mixture is spooned into a sausage stuffer and squeezed carefully into casings, keeping the amount even throughout. Take care not to introduce bubbles, which can turn your sausages into real bangers in the pan.

5 Once the sausages are linked they should be hung up to dry, ideally overnight.

Curing ham and bacon

This is another extremely useful skill to learn and can turn your pork into a much more valuable commodity. Curing has been used for generations the world over as a way of extending the shelf life of meat. Although fridges took away the need to preserve meat in this way, we still love the various types of ham and bacon and some products – particularly the air-dried hams favoured by the Italians and the Spanish – can command a handsome price.

Pork from any breed can be used for curing, though the traditional breeds are considered better because fat means flavour. Tamworths and Oxford sandy and blacks are often described as 'bacon pigs', though both are excellent dual-purpose pigs, producing excellent pork as well.

Basically there are two types of cure, wet and dry. The common ingredients in both processes are salt and patience. Curing isn't as difficult as some people may think, but it's something that can't be rushed and can involve a bit of trial and error. The first time you do it you may end up with something that isn't exactly to your taste, so practise with a small piece to avoid an expensive mistake.

You can make your own curing mixture or buy it ready-made from one of the many specialist suppliers. When you're just getting the hang of things, it's probably better to buy one of these tried-and-tested mixtures and then progress to your own blend once you've mastered the process. Almost all commercially prepared curing salts will include small quantities of sodium nitrate, which kill bacteria, help speed up the curing process and keep the meat pink. Saltpetre (potassium nitrate) was traditionally used in curing but, because it can be used to make explosives, it

Above: Dry-curing a piece of belly pork.

has now been banned in some countries and can be difficult to buy in others. If a recipe specifies both salt and saltpetre, just use a prepared curing salt in place of the two ingredients.

This isn't a cookery book, and as there are so many recipes to suit so many different tastes it's worth getting yourself a good book on curing (see Appendix 2) and experimenting until you find the mixture of ingredients that work for you.

Whatever you choose to cure with – with preservatives or without, organic or non-organic – the basic methods will be the same.

Below: Air-dried bacon which has been hung for over a year.

Dry curing

This is probably the easiest method to start off with. The example described here explains how to cure a piece of boned belly pork, which will give you streaky bacon, but you can use a boned loin for back bacon. To make things simpler, a commercial curing salt is used, along with an equal amount of brown sugar or molasses, a few crushed black peppercorns, and some crushed juniper berries. When you buy your curing salt it should come with instructions on how much to use according to the weight of the meat, but you might prefer to use your eye to tell you how much you'll need, based on your piece of meat.

1 Carefully bone the piece of belly pork.

2 Take the boned piece of belly and rub on the curing mixture, taking care to rub the meat all over and to work the mixture into every single nook and cranny. It's worth puncturing the skin side to allow greater absorption.

3 When you're satisfied the meat is completely covered, on every surface and in every possible area, place it in an airtight plastic container, ideally on top of some kind of platform – maybe another, smaller container like a butter tub, punctured with holes for drainage. Alternatively, you can vacuum-pack your meat and turn it daily. If you want a more intense flavour, rub in a little more of the mixture every day.

4 Allow a day per 2.5cm (inch) of depth of meat, eg 10cm = cure for four days, or leave for longer if you want a stronger, saltier flavour. Store in a cool place with a constant temperature during the curing process, such as a pantry or garage. You can also cure in a fridge, but you'll need to add a day or two to the curing time.

5 When you're ready, take the meat out and hang it in a cool place to dry a little before slicing. You can leave the meat hanging for as long as you wish, taking it down to slice as much as you need and then replacing it on the hook – as with Continental air-dried hams. Alternatively, slice and refrigerate for a more moist texture.

Air-dried hams

Much favoured by foodies across the world, air-dried hams can be an expensive delicacy. The key factors in curing in this way are temperature and humidity – which is why, traditionally, pigs would be killed when temperatures dropped, cured, and stored to provide a supply of meat through the winter. Spain has several regional versions of its jamón serrano ('mountain ham'); similarly, Italy has numerous variations of prosciutto (the most famous being prosciutto di Parma), while France has it jambon cru ('raw ham'). All are cured in much the same way and originated from the need to preserve meat for as long as possible. Most of these hams take a good nine months to two years to mature, and are eaten uncooked. Thin slices are taken off and the remainder of the joint is returned to its hanging place until next time. Ideally, debone your joint to make curing more straightforward.

2 When all parts of the meat have been thoroughly treated, place on a bed of curing salt in a plastic container.

1 Rub your curing salt mixture all over the meat as before, working it into any folds or crevices.

3 Cover with more salt. Seal the box and leave in a cool place for four to five days per kilo.

4 At the end of the allotted time, take out of the box, brush off excess salt, wrap in layers of muslin and hang in a cool, well-ventilated place. This stage of the curing process is the longest, taking at least nine months. Check every week or so to make sure the curing is working. White mould is fine, but green mould shows something has gone wrong.

Wet curing

This involves making up a brine solution in which the meat will be submerged until ready to cook. The brine can be used over and over again, provided it's kept in a cool place and sealed. For a basic, traditional brine solution, use the following:

- 4 litres (7 pints) water.
- 680g (1.5lb) curing salt.
- 680g (1.5lb) brown sugar.
- 10 peppercorns.
- 10 juniper berries.
- 1 bay leaf.
- 1 sprig of thyme or favourite herb.

Boil for five minutes and leave to cool. Strain through muslin to clear the brine.

To find out how much brine you're going to need, put your joint in the container you're intending to use and cover it with water, measuring how much you have to pour in. You'll need the same amount of brine.

Curing is easier and more consistent when the joint has been boned. However, if you want to cure on the bone care must be taken to get the brine solution right around the bone.

The best way to do this is with a brining pump. The needle of the pump has holes at intervals down its length and, when inserted deep into the meat, ensures the curing solution is evenly dispersed.

The joint can now be placed in the curing container and immersed in brine. The salt will cause the meat to bob to the top, so to keep it submerged you'll need to weigh it down with a plate or something sufficiently heavy – but not anything metallic, as the salt will corrode the metal.

The container should be placed in a cool room or in the fridge for three to four days per kilo and turned every day to ensure even curing. If you want a stronger, saltier cure leave it in a few days longer.

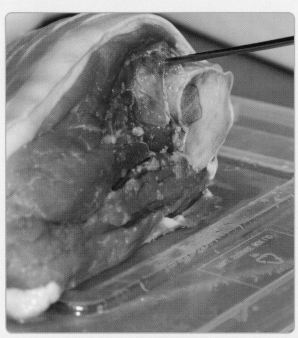

SELLING YOUR PRODUCE

Selling your produce

When you start raising your own pigs for meat it's only a matter of time before friends, neighbours or work colleagues start asking if they can buy some.

You will, of course, have to abide by the many rules and regulations regarding meat hygiene and the sale, storage and transport of food products, so, as advised earlier, do your homework and make sure you stay the right side of the law. Producing meat for your own consumption is one thing, but when you're putting animals into the public food chain you have to be extra-careful.

You may decide to go a step further than selling to people you know. Growing public concerns about traceability of food and animal welfare, coupled with an increased interest in buying local produce to support local businesses and reduce food miles, have helped farmers' markets to thrive.

Celebrity chefs have been great pioneers of local produce – particularly locally reared meat – urging TV viewers to buy food from their own communities.

There was a time when cheap meat was all that shoppers wanted. Indeed, some still do – and always will. However, more and more discerning buyers are asking where their meat came from, how it was raised, and how many miles it travelled before finding its way into our shopping baskets.

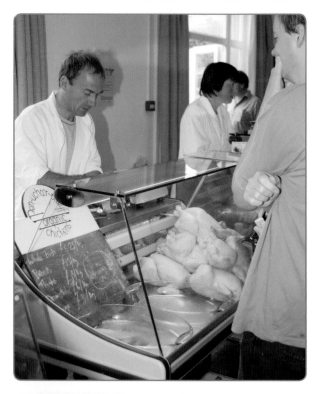

Above and left: Farmers' markets are a good way of selling direct to consumers.

Health and safety and the sale of food

There's enormous scope for disaster where food-handling is concerned, and numerous cases of illnesses caused by things like salmonella and E. coli have shown that no one can be too careful. Food poisoning won't just kill your business – at its worst, it could kill your customers too.

The Food Standards Agency (FSA) has prepared some excellent documents offering an introduction to the current legislation. These include FSA guidance on the requirements of food hygiene legislation and Starting up: Your first steps to running a catering business, which can be downloaded from their website. For more specific information about the type of business you're planning to run, contact your local authority's environmental health department.

Premises used for storing, preparing, distributing or selling food must be registered with your local authority at least 28 days before opening for business. Food businesses – whether they be shops, market stalls, mobile catering vans, vending machines or food delivery vans – are subject to the same regulations, and these regulations apply whether you sell the food publicly or privately, for profit or for fundraising. They do not apply to food cooked at home for private consumption.

Everyone involved in supplying food for sale for human consumption must meet basic hygiene requirements, in all aspects of their business. This covers everything from the premises and facilities used to the personal hygiene of staff.

Before you start any food-related venture, talk to the environmental health and trading standards departments at your local authority for advice. Make sure you understand the legislation, know what's required of you as a producer and seller of food, and be sure that you can fulfil all of the requirements.

You'll need to complete at least a basic food hygiene course at your local college, and possibly some more advanced courses, depending on what you have planned. All food businesses must have a Hazard Analysis and Critical Control Points (HACCP) manual. This is a food safety management system that identifies things which might go wrong in your processes, and details the measures you have in place to control or prevent hazards occurring.

All staff in contact with food must receive appropriate supervision, and be trained in food hygiene to enable them to handle food safely. There's no legal requirement for staff to undergo formal training, nor to have relevant qualifications, but many business operators prefer them to do so.

Left: Produce must be correctly labelled to comply with regulations.

Alamy

Selling at farmers' markets

The concept of farmers' markets originally came from the USA and is a fairly recent introduction to the UK. Bath became the pioneering city, setting one up in 1997, and today there are more than 550 operating throughout the UK, with more than half being members of the National Farmers' Retail & Markets Association (FARMA), the organisation set up to represent producers selling directly to the public. FARMA certifies farmers' markets in the UK that operate under its guidelines. Certification means they've been independently inspected and meet certain standards – stallholders must be local farmers, growers and food businesses selling their own produce.

Farmers' markets can be run by farmers' cooperatives, local authorities, community groups or private companies. Regulations vary from one market to another, but most are based on the same kind of criteria:

- Farmers' markets exist to enable local farmers and producers to sell direct to the public; to give consumers the chance to buy fresh, locally grown fruit and vegetables, locally reared meat and home-made products; and to raise public awareness on issues such as genetically modified (GM) foods and the importance of preserving the rural economy.
- Producers must have grown, raised, baked, processed or caught all food sold. The term 'producer' includes the stallholder's family and employees when they're directly involved in the business.
- Stallholders cannot sell products or produce on behalf of, or bought from, any other farm or supplier. This ensures complete traceability.
- Produce must be from a defined 'local' area. 'Local' is usually taken to mean within 30 to 50 miles of the market. However, producers from further afield may be considered if the produce they're selling cannot be sourced within the specified 'local' radius. In the case of applications for pitches by producers of similar foods, preference is normally given to the most local producer.
- The origin of the product, including where it was reared and/or processed, should be on all labelling.
- Stallholders must comply with all local and national laws and regulations regarding the production, labelling, display, storage and sale of goods. All producers must comply with the current food hygiene regulations (see the FSA website).

- Organisers will almost certainly ask to see the producer's public and product liability insurance certificate; producers will generally have £5 million worth of cover for each. Public liability insurance protects against claims by a third party injured or damaged as a result of your business, eg if your stall falls down and hurts someone. Product liability insurance protects against claims arising from the actual food you're providing.
- Producers should display trading names clearly on their stalls, together with a contact address.
- By law, prices must be clearly displayed, either on the pack or prominently on the stall. Having to ask for prices is off-putting to customers, and will scare some away.
- Most loose foods (eg fruit and vegetables) must be sold by net weight, using approved metric weighing equipment. If the food is pre-packed, the metric weight must be marked on the pack, but you can also add an imperial weight in a less prominent position.
- You must have good weighing scales, calibrated for metric weights and approved by your local Trading Standards officer. Spot-checks of your scales can be made at any time either at the market or at your farm.
- Producers 'adding value' to primary local produce (eg by curing or baking) should use local ingredients wherever possible. Some market organisers specify a minimum percentage of local ingredients to be used.
- Generally, products containing GM products are not permitted to be sold.

Farmers' markets – the pros and cons

The pros

A farmers' market is an excellent showcase for your produce, allowing you to sell quality food at a good price to shoppers who are happy to pay a little more for a premium product. Cutting out the middleman means you get the full retail price – sometimes as much as three times the wholesale price.

It can be very satisfying dealing direct with the public, talking about your produce, your farm and your lifestyle, and collecting feedback from satisfied customers. You don't get that when you hand over your produce to a middleman.

If you sell from your farm gate, customers may turn up sporadically. At farmers' markets, customers all turn up within the space of just a few hours.

Unlike opening a shop, the start-up costs are low. The main outlay (aside from the obvious costs involved with producing your food) involves renting a pitch and transport. Most transactions will be in cash, so you don't have to wait for cheques to clear.

The cons

You have to have commitment. Satisfied customers will expect to see you there every time, so it's no good deciding that you'd like next Sunday off. You'll disappoint potential repeat customers, who might spend their money elsewhere, and you'll also annoy the market organisers. You'll be expected to turn up regardless of the weather. Farmers' markets aren't a fair-weather occupation.

You must be prepared to put a phenomenal amount of time and planning into getting your goods ready for sale. You'll want to be selling fresh meat, rather than frozen, so you'll have to ensure that your animals will be ready for slaughter when you need them. This means working backwards to when your animals need to be born or bought in – which isn't all that easy – and you have to have a Plan B should you run short of stock.

Don't forget that you also have to allow time for butchery, processing, packing and labelling, probably just days before the market. And what happens if you don't sell everything on market day? What would you do with the leftover food? Customers generally prefer fresh produce; you'll need to think what you'll do with your leftover goods.

One step at a time

Draw up a business plan. Work out how much profit you'll need to make to support yourself and any partner or family, and how much food per week or month you'll need to produce and sell. Think about your setting-up costs: product and public liability insurance; buying and running a vehicle to transport your goods; buying, hiring, or building a stall from which to sell; buying storage facilities, eg a chiller and/or freezer; buying and/or designing packaging and labelling, etc.

Consider whether you have the time and commitment. Someone has to look after things at home on market day or while you're preparing for one. What would happen if you or your partner were to fall ill? Who would take over then?

Explore other ways of selling your produce. It may be that farm gate or mail-order and Internet sales would be less time- and energy-consuming and, potentially, more profitable.

Tips for success

If you really do want to give farmers' markets a try, you need to make sure your stall stands out from all the rest:

- Create an attractive and colourful display that will stop shoppers who are just browsing. Tell the story behind your produce. Include photographs and information about your livestock, and the food preparation process. If you've won any rosettes for your livestock or produce, show them off.
- First impressions are important, so get to the market in plenty of time to start setting up.
- Make sure all your products are clearly labelled, both to comply with legislation and to make life easier for your customers.
- Smile and be welcoming and think about your appearance. Crisp white overalls or colourful aprons always look smart and professional. Tie your hair back or wear a hat, and make sure your hands and nails are clean.
- Engage potential customers in conversation. Tell them about other farmers' markets you attend, offer samples, and give out flyers or business cards.
- Don't undervalue your products. Do your homework beforehand and find out what others supplying similar foods are charging.
- Use the media. Think of news angles to promote your business. Are you the first producer to be selling a certain type of food at your farmers' market? Are you using unusual ingredients or a novel way of processing? Has your business won a grant that's allowed you to boost production? Have you taken on new staff? Have you won an award for your produce? Have you won over a celebrity customer with your wares?
- Persevere! You may not be an instant hit at your first market, and it can take quite some time before you start building up regular customers who come and seek you out. However, if you stick at it, hang on to the same stall in the same place, and keep providing quality food, chances are you'll start to see the same faces coming back time and time again. The proof of the pudding is when new customers arrive saying friends have recommended your produce. That's the kind of thing that makes all the hard work and commitment worthwhile.

BREEDING

Breeding

Pig keepers become pig breeders for all sorts of reasons. On the one hand there are those who set out to do it and plan it all out; on the other there are those who just kind of fall into it.

Often what happens is that someone will buy two pigs, with the intention of filling the freezer, and then the inevitable happens – they grow too attached to them and decide to keep them to breed from. Or a mixed-sex batch of weaners mature a little faster than expected, and one thing leads to another, and before they know it brother and sister are destined to become mum and dad.

Anyone thinking of becoming a breeder should take a while to consider what might be involved:

■ Are you ready to become a pig breeder? Do you know enough about keeping pigs to take the next step? Pigs are fairly low-maintenance creatures – until you get to the breeding stage. Would you know what to do if something went wrong during farrowing?

■ Do you have the time and commitment? It's all very well keeping weaners to pork weight and postponing the annual holiday for four months until after they've gone to the abattoir, but now you're considering embarking on an all-year-round occupation; looking after pigs in good weather and bad.

■ Do you have enough land to support several fully grown pigs all year round, along with some of their offspring? It's all very well buying in two weaners at a time and keeping them until pork weight – that way you can choose to take a break and let your ground recover; but with full-grown sows – which can reach 400kg in weight – churning up and trampling your ground without respite you're going to need a lot more space. You'll need indoor accommodation or another, separate area for good ground when it comes to farrowing time.

■ What will you do with the piglets? Do you have a market lined up for them? Are you confident of finding outlets for the pork, or buyers for the weaners? A traditional breed sow can have anything between six to twelve piglets at a time; a modern or cross-bred one may have twice as many. At the lower end of the scale, two sows breeding twice a year might have 24 piglets between them; or at the higher end, that's 98 piglets to deal with.

These are just a few initial things to ponder. Graduating to pig breeder means stepping into a whole new world of responsibilities, so please think carefully before making the decision – or before allowing accidents to happen.

Left: Think about how many piglets you can cope with.

Selecting for breeding

Whether you decide to breed pedigree or not, the same basic rules apply when choosing stock. The best advice anyone can give is to buy your foundation stock from a reputable breeder. It's not a good idea to breed from pigs that you originally bought for meat. If they were sold as meat pigs, then that was for a reason – a reason that ruled them out as potential breeding stock. Good breeders operate on the basis of 'breed from the best, eat the rest', so remember this mantra.

A good pig should be strong and well nourished. Breeding stock should always be fed a nutritious diet, but should never be allowed to get overweight because that can not only cause health problems, but can also have a serious effect on fertility. Never breed from any pig which doesn't look fit and healthy, or which has genetic deformities.

A breeding pig should have good bone structure, a level back, strong, straight legs, and good depth of chest to allow plenty of space for the internal organs. Sows and gilts have to be sound enough to give birth to several litters, and they need to be solid enough to support the weight of a hefty boar – hence the importance of a good back and legs. The same goes for the boar, which has to be strong enough to do the job he's being reared for.

Above: A newborn piglet.

Below: Choosing a new boar.

Underlines

One of the main factors which eliminates a pig from being suitable for breeding is a bad 'underline'. This means the double row of teats, which should be in pairs and spaced evenly down the underbelly of the pig. This applies to both females and males, because either can pass on this important characteristic to their offspring. Rows of teats that are oddly spaced, missing or 'blind' (not fully developed and so non-functional) do no favours for the piglets when they come to suckle.

Insufficient numbers of teats, or ones that look like proper teats but don't work, mean that some piglets won't be able to feed well; they'll fall behind and may

even die. This is particularly crucial with modern breeds or cross-bred pigs, which can produce very large litters. There should be at least 12 teats, starting well forward, evenly spaced and in a straight line.

The front teats produce more milk than the back ones; the last two or three pairs can occasionally be wider apart, with the result they're often too tucked away under the sow's flank when she's lying down to feed her litter. This makes them less accessible, so sometimes they don't get used and simply dry up. Every teat has to work so that every piglet in the litter has a good chance of being able to suckle properly.

Below: Bad underline with misplaced teats.

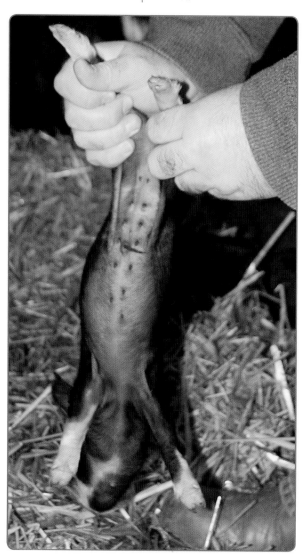

Below: Good underline – evenly spaced teats.

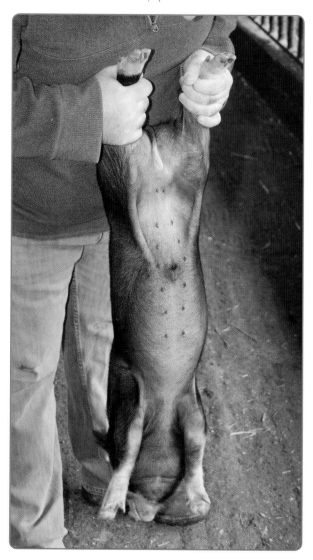

Buying pedigree breeding stock

Every breed has its own 'standard' – a checklist of what's required and desirable in a pig. If you're dealing with a reputable breeder, he or she may have done the selecting for you. No breeder with a valuable reputation will register as pedigree an animal unless it meets the breed standard. Every breed standard will demand a good underline, but it will also include specifications about anything from hair and skin colour to coat markings and the shape of head or nose.

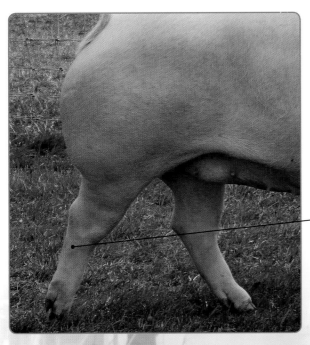

Breeders often describe a pig as being 'up on its pasterns'. This means the pig is walking well, appearing to be walking on tip-toe. An older or overweight pig will often be 'down on its pasterns' – with all parts of the foot touching the floor.

Solid and level back.

Good depth of chest.

General appearance: in a pedigree pig all the key characteristics meet the breed standard.

Well-nourished, but not overweight for age or condition (eg. if pregnant).

Deep well muscled hams.

Strong, straight legs.

Sound feet with strong, even, well-spaced toes which neither turn inwards nor outwards. Pig has a good gait, walking freely and easily, up on its pasterns. No sign of limping.

Good underline with the required number (usually 12, sometimes 14) of evenly spaced teats. Applies to males as well as females.

Healthy coat and skin with no sign of parasites.

105

Pedigree pigs – why bother?

So why is there so much fuss about keeping a pig with a piece of paper which states its ancestry? Surely you can just buy a couple of pigs and keep them and enjoy them, even breed from them, without going to the expense of getting into all that pedigree stuff? Well, yes, of course. But if you really love the breed and want to do what you can to help it survive, you should be buying animals that are registered as pedigree.

The best way to support rare breeds – whether pigs, sheep, cattle or whatever – is to help preserve the remaining bloodlines by producing good-quality stock. It's a sad fact of life that most of our native breeds have suffered a serious decline in numbers thanks to consumer demand for almost fat-free, tasteless meat. Many of the most recognisable breeds are still fighting for survival, with just a few hundred registered sows per breed.

Anyone wanting to register offspring from pedigree pigs (other than British lop and kunekune, which have their own registration systems) must be a member of the British Pig Association (see Appendix 1 for details). The Association holds records of pigs born as far back as 1884, when it was founded to preserve our native breeds, ensure traceability and uphold breeding standards.

Registration

The saying, 'A pig without a pedigree is just a pig' may seem a bit pompous on the face of it, but you have to understand that without that important piece of paper which bears the identities of your pigs, and details of their parents and grandparents, you have nothing to prove that they are what you say they are. They might look like Berkshires or Gloucestershire old spots, but for all anyone knows they could be mere mongrels which just happen to carry the dominant characteristics of one particular breed in their ancestry.

This becomes important when you start thinking about breeding. Unless your pigs are registered and offspring are birth notified and therefore eligible for registration, you won't be able to market any piglets as pedigree. Similarly, rare breed meat can attract a premium price and if you want to label your meat as pedigree produce you need the necessary paperwork, or you could end up in trouble with your local Trading Standards department.

Registered top-quality pigs are often bought by

Left: Pedigree pigs sold to others will bear your herd name and promote your stock.

smallholders who say they 'want to do something to help rare breeds'. Unfortunately, though, not everyone who buys registered pigs ends up breeding from them – which can be a real waste of good stock. Others may put their gilts or sows to a pedigree boar, but not birth notify or register the offspring. And then there are those who choose to cross-breed. The result in these cases is the same – on paper, those females have never been bred from and officially have no offspring, so their existence has done nothing to help their breed.

In some instances this can have serious implications, such as when a sow is from a particularly rare bloodline. With numbers of certain breeds so desperately low, we should do everything in our power to ensure the continuation of our indigenous breeds. Simply keeping a couple of pedigree pigs isn't helping your favourite rare breed. You may have temporarily helped a breeder a little by putting money in his or her pocket, but that's the only contribution you'll have made.

When you register as a pedigree breeder you need to choose a name for your herd, even if you only start off with one pig. This name is your prefix – it will be included in the pedigree name of every pig you register. Lots of people use the name of their farm as their prefix. There are three parts to a pedigree name: the herd prefix; the bloodline; and the pig's individual number. For instance, the Myfarmname herd of pedigree large blacks has a boar from the Defender bloodline that it wants to register. It's the 58th pig the herd has registered as pedigree, therefore the pedigree name is 'Myfarmname Defender 58'.

Pedigree identification

There are two ways of identifying pedigree pigs – tattooing and notching – and each breed society decides which method it prefers. In general, light-skinned breeds go for tattooing, while breeds with dark skins go for notching because tattoo inks don't show up as well. Some work has been done to develop white tattoo ink, but so far no one has managed to get it absolutely right. Instead, dark breeds have an ear notching system.

The diagram opposite shows how different locations on the ear correspond to different numbers. Notches must be inserted in the correct location to represent a pig's individual number. For example, two notches at the location 1 (units) would mean that pig was number '2'. Special notching pliers are available from farm suppliers. A V-shaped section must be taken out of the ear, and iodine is used immediately afterwards to prevent infection.

With tattooing, the BPA's basic requirement is for

Above: Tattooed ear with both HDL letters and numbers.

just the number to be tattooed, though lots of breeders include their herd designation letters (HDL) too. These are three letters given to you when you join, which act as a kind of shorthand identity on official paperwork.

Occasionally the tattoos fade or become illegible, but they can be replaced if you have written permission from the BPA.

Below: Notching may look complicated, but you will get the hang of it.

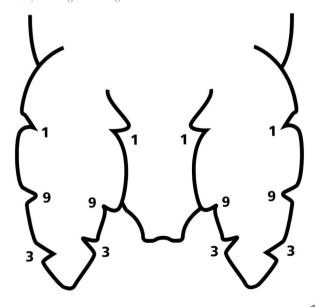

What age to breed?

A lot depends on how well grown a pig appears to be, but most good breeders will wait until pigs are between eight and ten months old before mating. This applies to both gilts and boars. A gilt may come into season for the first time at five months, but that doesn't mean she's physically prepared to sustain a successful, productive pregnancy. Similarly, a boar will start producing sperm around four months old, but studies have shown that semen collected will contain a lot of immature and abnormal sperm.

Care should be taken in matching a boar to a gilt, because a boar which is too heavy can seriously injure a mismatched gilt. The boar's weight can cause her to collapse, possibly splaying her legs, and internal injuries can also be caused.

Put two virgins together and they may fumble about a bit and get nowhere – or even injure themselves in the process. A young boar paired with a maiden gilt will need to be supervised at the first mating and literally given 'a helping hand'. An inexperienced boar may not get it right first time if left to his own devices. Valuable time can be lost, with the result that you may need to wait until the gilt's next season, and the boar's confidence can be knocked if it keeps slipping off and not getting anywhere. Guide his penis into the

Above: Piglets just hours after being born.

vagina and make sure it locks into the cervix. If semen leaks out, it isn't in far enough.

Never breed from a pig that's showing signs of illness, lameness or injury. Make sure that the boar and the sows or gilts are up to date with their vaccinations to ensure immunity is passed on to their offspring.

Below: A young boar gets a 'helping hand' to mate a maiden gilt.

The breeding cycle

THE BREEDING CYCLE

In order to breed successfully, you need to fully understand the breeding cycle. In the wild a pig would have only one litter a year, weaning the piglets off her milk and on to solid food when she felt ready. However, most breeders aim for two litters a year, with commercial producers striving for 2.5. The latter is achieved by early weaning, at about three weeks old.

The well-known approximation of how long a pig is pregnant for is 'three months, three weeks, three days'. It's usually accepted to be 114 days, but can vary from 112 to 115, depending on the breed.

A gestation chart like the one shown here will give you a guide as to when piglets will be born. Note the date of service on the top line and the delivery date will be immediately below.

Served	January	1	2	3	4	5	6	7	8	9	10	11	12	13	14	15	16	17	18	19	20	21	22	23	24	25	26	27	28	29	30	31
Due	April/May	26	27	28	29	30	1	2	3	4	5	6	7	8	9	10	11	12	13	14	15	16	17	18	19	20	21	22	23	24	25	26
Served	February	1	2	3	4	5	6	7	8	9	10	11	12	13	14	15	16	17	18	19	20	21	22	23	24	25	26	27	28			
Due	May/June	27	28	29	30	31	1	2	3	4	5	6	7	8	9	10	11	12	13	14	15	16	17	18	19	20	21	22	23			
Served	March	1	2	3	4	5	6	7	8	9	10	11	12	13	14	15	16	17	18	19	20	21	22	23	24	25	26	27	28	29	30	31
Due	June/July	24	25	26	27	28	29	30	1	2	3	4	5	6	7	8	9	10	11	12	13	14	15	16	17	18	19	20	21	22	23	24
Served	April	1	2	3	4	5	6	7	8	9	10	11	12	13	14	15	16	17	18	19	20	21	22	23	24	25	26	27	28	29	30	
Due	July/August	25	26	27	28	29	30	31	1	2	3	4	5	6	7	8	9	10	11	12	13	14	15	16	17	18	19	20	21	22	23	
Served	May	1	2	3	4	5	6	7	8	9	10	11	12	13	14	15	16	17	18	19	20	21	22	23	24	25	26	27	28	29	30	31
Due	August/Sept	24	25	26	27	28	29	30	31	1	2	3	4	5	6	7	8	9	10	11	12	13	14	15	16	17	18	19	20	21	22	23
Served	June	1	2	3	4	5	6	7	8	9	10	11	12	13	14	15	16	17	18	19	20	21	22	23	24	25	26	27	28	29	30	
Due	Sept/Oct	24	25	26	27	28	29	30	1	2	3	4	5	6	7	8	9	10	11	12	13	14	15	16	17	18	19	20	21	22	23	
Served	July	1	2	3	4	5	6	7	8	9	10	11	12	13	14	15	16	17	18	19	20	21	22	23	24	25	26	27	28	29	30	31
Due	Oct/Nov	24	25	26	27	28	29	30	31	1	2	3	4	5	6	7	8	9	10	11	12	13	14	15	16	17	18	19	20	21	22	23
Served	August	1	2	3	4	5	6	7	8	9	10	11	12	13	14	15	16	17	18	19	20	21	22	23	24	25	26	27	28	29	30	31
Due	Nov/Dec	24	25	26	27	28	29	30	1	2	3	4	5	6	7	8	9	10	11	12	13	14	15	16	17	18	19	20	21	22	23	24
Served	September	1	2	3	4	5	6	7	8	9	10	11	12	13	14	15	16	17	18	19	20	21	22	23	24	25	26	27	28	29	30	
Due	Dec/Jan	25	26	27	28	29	30	31	1	2	3	4	5	6	7	8	9	10	11	12	13	14	15	16	17	18	19	20	21	22	23	
Served	October	1	2	3	4	5	6	7	8	9	10	11	12	13	14	15	16	17	18	19	20	21	22	23	24	25	26	27	28	29	30	31
Due	Jan/Feb	24	25	26	27	28	29	30	31	1	2	3	4	5	6	7	8	9	10	11	12	13	14	15	16	17	18	19	20	21	22	23
Served	November	1	2	3	4	5	6	7	8	9	10	11	12	13	14	15	16	17	18	19	20	21	22	23	24	25	26	27	28	29	30	
Due	Feb/March	24	25	26	27	28	1	2	3	4	5	6	7	8	9	10	11	12	13	14	15	16	17	18	19	20	21	22	23	24	25	
Served	December	1	2	3	4	5	6	7	8	9	10	11	12	13	14	15	16	17	18	19	20	21	22	23	24	25	26	27	28	29	30	31
Due	March/April	26	27	28	29	30	31	1	2	3	4	5	6	7	8	9	10	11	12	13	14	15	16	17	18	19	20	21	22	23	24	25

Getting the timing right

Oestrus is the period during which a sow or gilt ovulates and is receptive to the boar. When a pig is 'in oestrus' she's also described as being 'in season' or 'on heat'. This happens approximately once every 21 days, though with gilts it can be once every 18 to 24 days. The signs can vary between pigs, and are easier to recognise in sows than in gilts. There will be behavioural changes – pigs can lose their appetites and become restless and 'needy', wanting to be fussed over and petted. They may become much more vocal, growling and squealing for attention. Some will sniff the genitals of their female pen mates or attempt to mount them.

The first physical sign of the onset of oestrus is a swelling and reddening of the vulva. A small amount of mucous may also be visible. This stage is known as proestrus; the sow or gilt may act as if she's ready for the boar, but she probably won't stand to be served.

Around 12 to 24 hours later the vulva will be less swollen and paler, and the mucous will have thinned to a clear fluid. By this time she should stand rock solid if pressure is applied to her back, showing she's ready to be served. This is known as 'standing heat'. If you sat on the back of a sow that was in full heat, she wouldn't walk away. The reaction will last anything between 10 and 20 minutes, during which she will stay motionless, with her back slightly arched and a slightly dazed look in her eyes. Her tail may be lifted to one side and her ears may be cocked or turned back.

Below: A swollen vulva is a good sign of the onset of oestrus.

Mating period

Placing a female pig within sight, sound and smell of the boar – but not in the same pen – will make oestrus easier to detect. The boar will become extremely 'chatty' and excited, pacing up and down the perimeter and frothing at the mouth.

Aerosol sprays have been created which replicate the smell of a boar. All you need to do is squirt some on a cloth and hold it to the pig's nose. Assessing when your pigs are ready to be served becomes even more important when attempting to impregnate by means of artificial insemination (more of this later), because timing is crucial to success. It's important to remember that showing signs of oestrus – in the ways described above – does not mean she's ready to be mated or inseminated.

Between 38 and 42 hours after the start of oestrus, ovulation occurs. The process takes around four hours. The lifespan of good-quality sperm is around 24 hours – twice that of the egg, which is only really at its best for six to eight hours after ovulation. Therefore the best time for insemination, whether natural or artificial, is shortly before the release of the eggs. Because of the difficulty of identifying the timing of ovulation with any accuracy it's best that mating takes place at least twice during the time when a pig is in standing heat. Sows and gilts will normally be served several times during the 48 hours or so that standing heat lasts.

Above: Artificial pheromone sprays can help detect whether a female is ready.

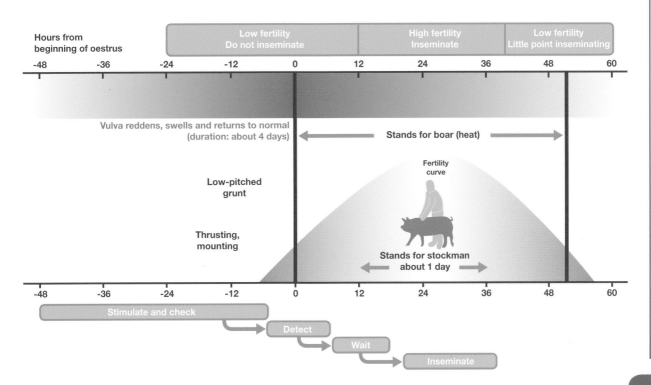

Some breeders will put a boar together with sows or gilts as soon as signs of standing heat occur and leave them together until the optimum time for impregnation has passed. Others separate them after the first service and reunite them at 24-hour intervals until standing heat is over, saying this makes for a better mating.

Whether you take the boar to the females or the females to the boar is another debatable point. Some say that if the boar is taken to the females, he may be distracted by the sight and smell of a new pen and so waste time. Therefore their choice would be for the females to visit the boar's abode.

Others prefer for the boar to go to the females. Stud boars often go out on hire to protect the owner's stock from infection by visiting females. The boar is normally kept for two cycles, to check whether impregnation has occurred and, if not, to allow a second attempt.

A further argument for this system is that moving a sow or gilt after mating can adversely affect implantation of fertilised eggs. Experts advise that it's not safe to change the environment of the sow in any way between day three post-service and day 30 post-service.

For these reasons, sows and gilts should be kept quiet and unstressed after being served, and no major changes should occur – ie don't move them to new quarters; don't mix them with unfamiliar pigs, in order to avoid fighting and stress; don't think of transporting them anywhere; and avoid giving vaccinations or carrying out other traumatic procedures.

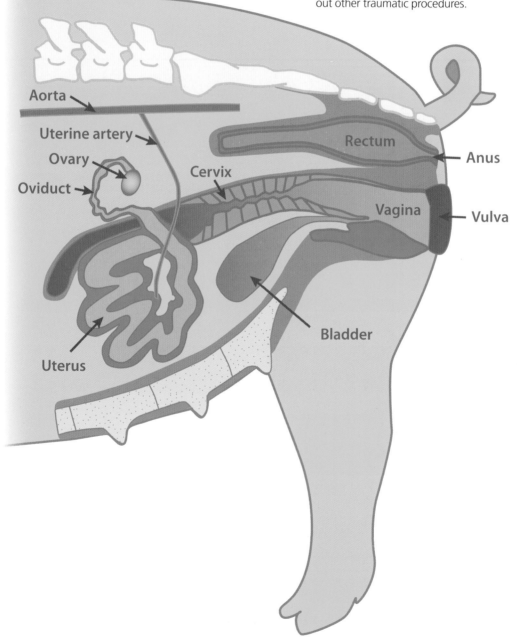

Aorta

Uterine artery

Ovary

Oviduct

Cervix

Rectum

Anus

Vagina

Vulva

Bladder

Uterus

Keeping a boar

If you intend only keeping two sows, it will probably not be worth your while having your own boar. Unless you're going to hire him out he's likely to get bored and restless, and his sperm may get stale and lose fertility.

With just two females your boar is only going to be put to work four times a year. Also, once you've bred successfully from him you'll not be able to mate him with his own offspring. In addition there's the cost involved in keeping him all year round; an adult boar will be eating around 3kg or more food a day, so you have to weigh this up against his value to you.

How long does mating take?

Mating can take anything from 10 to 20 minutes. It's not unusual to see a boar appear to fall asleep on the back of a sow or gilt; he has around three quarters of a litre of semen to deliver, after all.

With rare breeds, finding a suitable boar may be difficult without travelling long distances, so keeping your own boar might be the only option. If you decide to do this, make sure your fencing is up to the challenge, otherwise every time your females come into season you could have difficulty containing him.

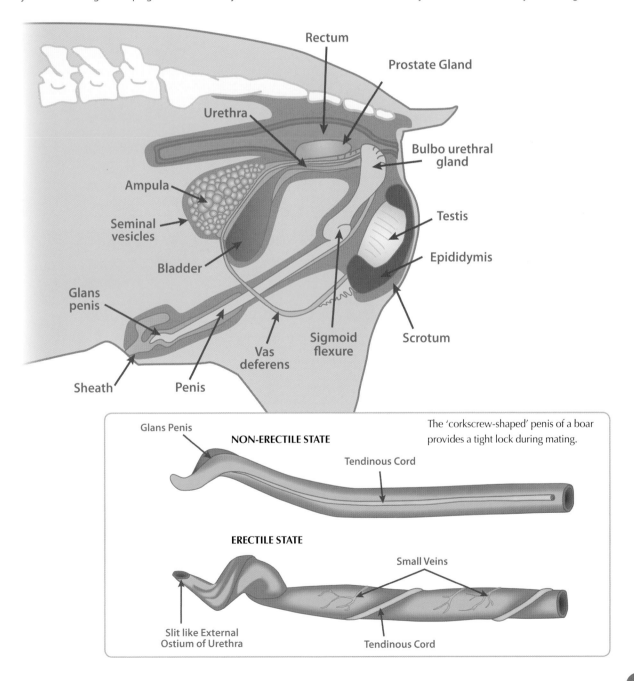

Rectum
Prostate Gland
Urethra
Bulbo urethral gland
Ampula
Seminal vesicles
Testis
Bladder
Epididymis
Glans penis
Sheath
Penis
Vas deferens
Sigmoid flexure
Scrotum

Glans Penis
NON-ERECTILE STATE
Tendinous Cord
The 'corkscrew-shaped' penis of a boar provides a tight lock during mating.

ERECTILE STATE
Small Veins
Slit like External Ostium of Urethra
Tendinous Cord

Artificial insemination

An alternative to using a boar is artificial insemination (AI). This can also be a useful way of getting hold of a different bloodline, reduces the risk of introducing disease to your herd, and saves on hiring in a boar, paying stud fees and feeding him while he's with you.

There are numerous AI companies providing semen, but the only one that keeps any great number of rare breed boars at stud is Deerpark Pedigree Pigs in Northern Ireland (see Appendix 1 for details).

One drawback is that AI isn't as easy with gilts as with sows, and most breeders recommend serving with a boar for the first litter. Part of the problem with gilts is that it isn't as simple to spot oestrus, the single most important factor with AI.

If you know your pig is showing signs of oestrus, you can order by phone and have your semen delivered the next day. Three bottles or flat-pack sachets will be sent (you need to specify which you prefer), along with three catheters that are used for inserting the semen deep into the cervix. You'll have a choice of catheters: one with a fat bullet-shaped head for sows; a similar, but smaller one for gilts; and one with a spiral head which mimics the appearance of the boar's penis and 'locks' into place in the cervix.

Care of the semen

The semen arrives in a polystyrene box, and should be kept in the box until it's ready to be used. The ideal temperature for storing semen is 17°C, but it can be kept in a room between 15°C and 20°C. Whatever you do, don't put it in the fridge, because it will be much too cold and the semen will be damaged. Similarly, don't expose the bottles or packets to sunlight, as UV rays can destroy sperm. AI companies dilute the semen with a special chemical that extends its 'shelf life' to between three and five days. Beyond this the viability will be doubtful. You'll need to keep turning the bottles or packets occasionally in order to mix the semen and other contents.

Below: A round-tipped foam catheter is straightforward to use.

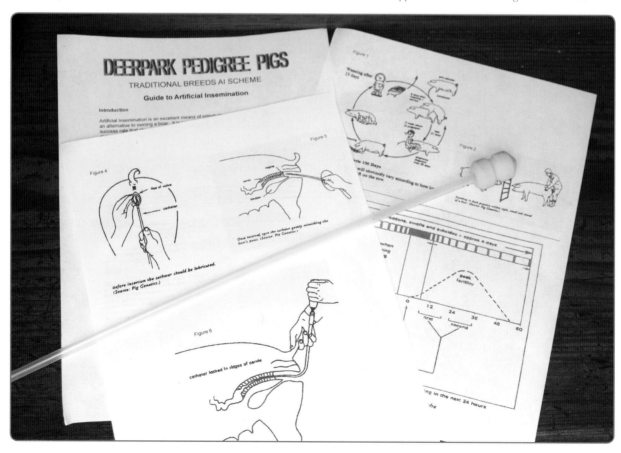

Pre-ovulation nutrition

A dry sow or gilt at breeding age will generally be fed around 2kg to 2.5kg a day, depending on breed and condition. However, in preparation for mating the ration can be doubled 10 to 14 days before insemination, in order to improve ovulation. The theory is that the body recognises there's a plentiful supply of food and knows there will be sufficient nourishment for future embryos. Following mating, feed should be gradually reduced back to the normal level for the majority of the pregnancy.

A pig shouldn't be too fat at the time of service, as it can affect fertility and cause small litters. This is because the ovaries can be encased in fat and therefore the eggs can't find their way into the fallopian tubes. Also, if she's fat at the time of farrowing, that, too, can cause complications, as layers of fat in the cervix can make it difficult for large piglets to emerge. A fat pig is likely to have a slow, inefficient farrowing.

A further adjustment is needed in the days leading up to farrowing. Rations should be cut by one-third two days before farrowing, and then gradually increased afterwards (as explained later) as the piglets grow.

Don't neglect the boar either. As the saying goes, 'the boar is half of your herd', so treat him accordingly. Ensure he's well nourished before mating, to allow for maximum sperm production, and to maintain his stamina.

Below: These gilts, approaching breeding age, are in the peak of condition.

Insemination procedure

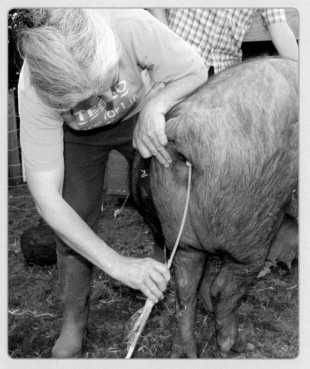

1 First establish that your pig is 'standing', as explained earlier. Wait 24 hours, test her again and, if she stands, inseminate.

2 Warm the semen container in your pocket.

3 You may need to act like a boar to help keep her standing: a boar's 'foreplay' involves giving off pheromones, nudging the female in the belly and flank, and stimulating her vulva by sniffing and licking. Only then will he attempt to mount her. If you don't have a boar within easy reach, use some boar spray to help get her in the mood and try simulating what the boar would do before finally putting pressure on her back and seeing if she stands solid.

4 Clean the vulva with water and a paper tissue. Do not use soap or disinfectant, as these can affect the viability of the sperm.

5 Apply a lubricant to the end of the catheter.

6 Hold the sow's tail with middle, fourth and little fingers, and use the thumb and index finger of the same hand to open the vulva.

Angle of the catheter

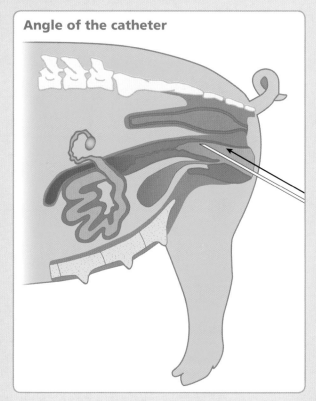

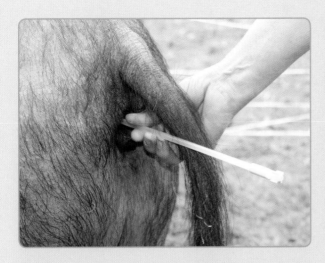

7 Hold the catheter at an angle of 45° and gently but firmly push upwards to ensure it doesn't enter the bladder.

8 When the catheter reaches the entrance to the cervix an obstruction will be felt. What happens next depends on what kind of catheter you're using. With a spiral catheter, turn in an anticlockwise direction (towards the left side of the sow) until it's locked in the cervix. With a rounded catheter, gently push forward and the ridge in the sponge end should slot into place in the cervix, so that you feel a resistance if you try to withdraw it. In both cases this lock creates a seal, preventing the semen from flowing out of the vulva.

9 Take the insemination bottle or packet from your pocket and attach it to the catheter. Apply gentle pressure until all the semen is discharged into the uterus. If the semen starts leaking back out of the vulva, stop inseminating and try to reinsert the catheter with a better lock. The whole process could take 10 minutes or more, so be patient. Don't try and squeeze it in a little faster. Just let the uterine contractions work at their own rate. Remember that a boar could take 20 minutes to do this job! If you can, keep stimulating the sow to encourage uterine contractions, which will help the semen along its way.

10 When the final drops of semen have disappeared from the container, wait a few minutes, then remove the bottle or fold down the sachet, leaving the catheter inserted. Modern catheters have a plug attached that can be inserted in the end of the tube to prevent semen coming back out. After insemination a boar will ejaculate a jelly-like plug which does the same job. A spiral catheter should be removed by turning gently in a clockwise direction. A sponge catheter can be left in until the pig releases it naturally – but don't forget to collect it later.

11 Further services are normally carried out once or twice a day until standing heat has passed. Remember the aftercare advice given earlier: keep the pig in as restful an environment as possible.

12 Any unused semen should be disposed of. Don't keep it for the next time a sow needs serving, as it probably won't be viable.

13 Note the first date of service and check for signs of pregnancy or wait to see if the sow or gilt returns to service (comes back into season) in 21 days' time. If this doesn't happen you can assume that you've been successful.

Pregnancy diagnosis

Keep records of when your pigs are in oestrus and when they're served. Check 21 days after service to see if they've 'returned', ie come back into season. If not, it's likely they're in pig.

The easiest and most natural way of checking to see if a pig is pregnant is to introduce her to a boar as described earlier when discussing standing heat. Alternatively, you can use an ultrasound pregnancy tester, like this Rotech Döppler (pictured). The probe, placed on the udder near the flank, passes high-frequency sound waves into the sow's abdomen. Sounds produced by blood passing through the uterine artery can be picked up from 25 days after service to the end of the pregnancy. They can also be used during farrowing, to detect whether there are any more piglets inside the sow.

It is also possible to detect pregnancy by manual examination. This involves clearing the rectum of faeces and feeling for a vibration – known as 'fremitus' – in the middle uterine artery. It's often described as a 'thrilling pulse', rather like a mild electric shock. However, this technique is almost exclusively performed by experienced veterinarians, and should not be attempted without expert supervision.

Above: Are they or aren't they?

Below: Pregnancy testing using a Rotech Döppler.

Farrowing

Preparation

Whether you farrow indoors or outdoors is up to you. Allowing sows to farrow in the familiar surroundings of their own arks and paddocks may seem more natural, but there's a lot to be said for the controlled atmosphere of a barn, where monitoring the sow and litter is a whole lot easier. And, should veterinary attention be required, the sow is in a confined space and more easily accessible.

Wherever you decide to let your sows farrow, use the month or so before their litters arrive to carry out any vaccinations which may be needed. Sows vaccinated at this stage are able to pass immunity to their offspring via the colostrums they produce. This inherited immunity to certain infectious diseases can last five or six months. Consult your vet for advice on what should be done at this time.

Above: Newborn piglets will benefit from an additional heat source.

Outdoor farrowing

If you're farrowing outdoors, a small run should be made around the sow's ark to protect her and her litter from others in the herd. A low fence should also be erected around the entrance to the ark to prevent the piglets from getting out while still allowing the sow freedom to leave the farrowing area and wander around as she normally would.

If possible, rails should be fixed along the insides of the ark to prevent the sow rolling against the walls and unwittingly squashing a piglet. This is one of the commonest causes of dead piglets. If the ark is large enough, a 'creep' area should be constructed – an area separated by a sheep hurdle or some other pig-proof barrier which will be a place where piglet creep food can be placed, out of the reach of the mother.

Allow plenty of clean bedding, but not so much that newborn piglets can get lost while struggling to find their way to the nearest teat.

Supervising a farrowing in an ark isn't the easiest of jobs because of the lack of space, so if you're going to leave your pig to farrow alone, taking extra precautions for the safety of the litter makes perfect sense.

Indoor farrowing

Pens that have previously been used should be thoroughly cleaned using a pressure washer and thoroughly disinfected. The pens should be allowed to dry before sows are placed in them. As with outdoor farrowing arks, it helps to have

rails around the perimeter to minimise the risk of sows trapping piglets against walls.

Pens should, if possible, have a creep area available – either a corner of the pen fenced off with a hurdle, or maybe half of a sheep ring feeder – with a heat lamp suspended over it. This will serve as a safe and warm place for the piglets to sleep. Without such an area, piglets will cuddle up to their mother for warmth, or lie in her warm bedding once she gets up, and risk getting flattened she when flops back down or rolls over.

Bottom: A 'creep' area will give piglets their own safe space to sleep.

Getting sows ready

Sows should be moved to their farrowing quarters at least a week before they're due – sooner if you're not sure of the service dates. Gilts may need a little longer to get used to their surroundings, so you might want to move them two weeks before they're due. You should use this time to 'bond' with your pigs, particularly the first-timers. Gilts can sometimes need a great deal of reassurance during farrowing, so building up a close and trusting relationship is essential.

Reduce the feed ration gradually in the final week of pregnancy, so that you're giving just a third of the normal amount in the 48 hours before farrowing. By this stage the pig will be much less active and may go off her food a little. Reducing her feed will help ensure that she doesn't produce too much milk too soon and so is at less risk of getting mastitis (see Chapter 5). It also prevents her from getting bloated and leaves a little more space inside as her body enters the final stages of pregnancy.

Above: A newborn Tamworth with umbilical cord still attached.

Signs of the onset of farrowing

Pigs can react in a number of ways as farrowing approaches. Around 10 to 14 days before farrowing, the mammary glands and the vulva begin to swell. Teats grow in size and the veins supplying the udder become prominent.

Closer to farrowing the udder drops further, swells and tightens, the nipples darken and become more prominent, and there's an increase in pulse and respiration rates. Around 48 to 24 hours before farrowing, drops of milk can be expressed from the teats, but around eight hours before larger quantities can be expressed with ease. This is one of the best indicators that farrowing is about to start.

Below: A Berkshire sow and her piglets in the farrowing nest.

Above: Sows may continually re-arrange their bedding before and after farrowing.

Nesting

Whether the sow or gilt is to farrow outdoors or inside, she'll build a nest, anything from a week to a few hours before giving birth. It's a fascinating process to watch. If outdoors, she'll collect twigs, leaves, grass and even old feed sacks – anything that's moveable – and take it to her chosen farrowing place. She'll scrape everything into the desired shape before settling down. If due to farrow in a pen indoors the pig will exhibit the same behaviour, collecting all the available straw and constructing a nest in one corner, carefully building a 'wall' around herself as if she were preparing to protect her litter against the elements.

The birth – what to expect

Unfortunately, there's no blueprint for a farrowing. Each one is different, just as each pig is different. Most farrowings are trouble-free and piglets may be born before you realise your pigs are ready to give birth. It may be some comfort to know that statistics say the sow will manage 99% of farrowings on her own. Having said that, however, complications can arise and it's worth being prepared.

Behaviour changes

In the hours leading up to farrowing the pig will become restless, repeatedly lying down and getting up again,

making a lot of noise and maybe urinating frequently. Eventually, as the contractions increase, most will lie down and go into a kind of half-asleep state, breathing more deeply and giving low grunts as the piglets begin to appear. Often the sow will not stir until all the piglets have been born.

Others, however – particularly first-time mothers – appear extremely disturbed by the whole experience and find it impossible to settle. They can pace around, tearing at walls, biting metal bars, chewing concrete, and they occasionally give birth while they're on the move.

Agitated gilts are often at risk of savaging their newborns. Some will just lunge at piglets, in an effort to drive them away, while others will bite them, causing deep and often fatal wounds, or toss them across the pen, killing them in the process. In some cases the piglets will be eaten.

No one knows exactly why they do it, but it's linked to the major hormone changes which occur around the time

Be prepared

FARROWING CHECKLIST

- A thermometer (a digital one is easier to read).
- Rubber gloves or sleeves.
- Obstetrical lubricant.
- Towels to dry the newborn piglets as they arrive.
- A mixture of tincture of iodine and surgical spirit to treat the umbilical cords.
- A large box filled with straw or towels, in case you need to remove the piglets while the sow continues farrowing.
- Antiseptic spray, for treating injuries caused by a clumsy mother accidentally treading on a piglet or savaging it.
- Sedative (prescription only) in case first-litter gilts become agitated or aggressive.
- Oxytocin (prescription only) in case of delays in farrowing or problems with lack of milk (agalactia).
- Your vet's telephone number.
- Patience and a calm attitude. You can't rush this!

of birth. It could be that the sows associate the arrival of the piglets with the pain of the contractions. Or it may be that the sow, in such a highly-strung and agitated state, is just so completely freaked out at the sight of another living thing in her pen that she lashes out in the same way as a sow would normally react should a strange piglet wander too close to her own.

Vets can prescribe a sedative, azaperone (Stresnil), to calm the sow's behaviour. It normally takes at least 15 minutes to work. It's also used when difficult examinations have to be carried out, or to reduce aggression when groups are mixed.

With or without drugs, the key is to get the pig to lie down and complete the farrowing. This may be achieved by stroking her and talking to her, but a lot depends on your relationship with her. Getting to know your pig and developing a relationship of trust is essential in the run-up to farrowing, and occurrences like this prove the point.

All piglets should be removed to a safe place (the creep area or a box or bucket in a warm place) until the sow calms down and farrowing is completed. Piglets can be introduced to suckle, one by one, with the owner keeping a close eye on the reaction of the mother. Normally, once the piglets are suckling the mother will accept them – but be on guard just in case you have an exception.

Some breeders will cull gilts that display savaging tendencies as soon as they've weaned their litters. Others believe in giving a second chance, as bad first-time mothers are often absolutely fine second time around.

Below: Newborns receive no help from the sow on arrival.

Parturition

Piglets may be born head-first or tail-first, and neither position appears to make any difference to how easy or difficult the birth is. Most of the time the foetal sacs are broken at delivery, but some piglets may be born with the membrane intact, covering the head, and will die of suffocation without assistance.

Watch the sow for signs of each piglet's arrival. She may shiver, twitch her tail and raise her hind leg as the contractions propel the piglet down the birth canal. As the piglets are located in two different horns, the sow may turn over between each birth – or she may just stay on the same side throughout. No two sows are the same.

A normal farrowing can take anything between one hour and eight. Births should ideally be around 15 to 20 minutes apart.

As each piglet is born, check to see that its mouth and nose are clear of mucus, dry it with a towel, and spray its umbilical cord with a half-and-half mixture of iodine and surgical spirit to stop infection and to aid drying. Place the piglet at the sow's udder and help it to latch on to a teat. If your presence is upsetting the sow, leave her and only intervene if she or one of her piglets is in difficulty.

If the interval between the first and second piglet is more than 30 minutes, the second is more likely to be stillborn. Ask your own vet for advice, but most will agree that if a gap of an hour has passed and the sow is still straining, you should intervene. There could be a number of reasons for the delay:

1 There could be large piglets and a small pelvis.
2 The sow could have a lot of internal fat blocking the way.
3 Two piglets from different horns may be stuck in the birth canal.
4 Dead piglets may be causing a blockage.
5 There could be mummified piglets.
6 The womb may not be contracting as it should.

Sows that have been continually straining can suffer from uterine inertia. The muscles of the womb can stop contracting, leaving piglets queued up just beyond the cervix. An injection of oxytocin can kick-start contractions again, but you first have to check whether anything is causing an obstruction. Often the introduction of an arm into the vagina will stimulate activity, as can piglets suckling, or massaging of the udder.

If you do need to carry out an internal investigation, have a bucket of clean warm water ready containing a mild antiseptic and an obstetrical lubricant. Never use detergents because they can irritate, and never try and insert an unlubricated arm into a sow. If you have a plastic arm sleeve, use it to reduce the risk of contamination. Otherwise, clean your hand and arm thoroughly.

As you put your hand into the vagina, keep your fingers together. Push gently forwards until you reach the cervix and can feel the entrance to both horns of the womb. If you've

Farrowing

1 The first piglet arrives.

2 The piglet has to find its own way to the udder.

3 There is a risk of hypothermia if piglets get stranded away from the warmth of the sow.

4 When farrowing is completed, the placenta is shed.

5 Two hours later: a healthy litter of piglets, suckling contentedly.

Abortion

Pregnancies are frequently lost between 70 and 100 days and can occur for various reasons, including disease, extreme dietary deficiencies, parasites or genetic defects in the litter.

Embryos that die within the first six weeks of gestation are reabsorbed. Foetal remains can sometimes be seen in the afterbirth as white specks, which are bits of calcified skeleton. Those that die during the later stages are retained and expelled as mummified foetuses at the same time as the rest of the litter.

Above: Piglets and mum sleeping soundly at the end of farrowing.

delivered lambs you'll be used to going in as far as the elbow, but with pigs you need to be in as far as your armpit, so keep going.

Feel how they're presented and ease them out accordingly: if a piglet is coming head-first, place your hand over the head and hook your index and middle fingers round the nape of the neck. Should the piglet be in a breech or backward position, hold the hind legs and ease it out. Sometimes piglets might appear too large to deliver, but it can usually be managed with the help of a lot of lubrication and patience. However, if you don't feel sufficiently confident to cope, call your vet. Don't just leave the sow to strain and hope that everything will be okay. It probably won't.

If you do have to carry out an internal examination, give the pig an injection of a suitable antibiotic to guard against potential infection.

Reviving piglets

Some piglets, particularly those delivered by hand, are often suffering from lack of oxygen. If a piglet arrives and appears not to be breathing, clear the nose and mouth and stick a piece of straw up its nose. There may be mucus clogging the windpipe, and a coughing reaction might dislodge it. Another method is to take the piglet by the back legs and, with arms extended down, swing it from one side to the other, again to shift any mucus from the windpipe and the back of the throat.

Looking after newborn piglets

More than half of the piglets that die before weaning do so in the first 72 hours after birth. They lose heat quickly, so it's really important to keep them dry, warm and clear of draughts. Sows will not fuss over a piglet. Ewes and cows pay great attention to their newborns, licking them dry and bonding right from the start, but a sow hardly seems to notice when a piglet is born. She'll just lie there and wait for the others to arrive, maybe get up to eat or drink when it's all over, and eventually fall asleep, leaving her offspring to fend for themselves. More piglets die of hypothermia in the first few hours after birth than from any other cause.

Dead piglets

One of the most disappointing sights at farrowing is the discovery of a large, well-developed piglet lying dead in the pen. Unless you've observed each piglet being born you probably won't be able to tell whether it was born alive or dead.

Often during over-long farrowings an unborn piglet's umbilical cord can break or become pinched or stretched, cutting off its supply of oxygen. Piglets can suffer irreversible brain damage within minutes and then die – hence the need to intervene quickly if you think a piglet might be blocking the birth canal and holding up farrowing.

Piglets born fit and healthy are frequently crushed to death by their mothers. Sows can lie on top of piglets, tread on them, or squash them against the wall as they turn over. The piglet is programmed to squeal when it feels pressure, but not all sows are quick enough to respond. Other piglets found dead include those that didn't make it to a teat and died of hypothermia.

To determine whether the piglet was born alive, carry out a mini post-mortem examination, taking sample sections from the lungs. If lungs float, the piglet had inhaled air and was alive (the remaining air gives the tissue buoyancy). If they sink, the piglet was stillborn, as the lungs didn't contain air. The lungs of a stillborn piglet will also be darker in colour.

Above: Day-olds suckling contentedly.

Below: Stillborn or not? Left: this piglet was born alive but died later – see pink lungs. Right: this piglet, born dead, has much darker lungs.

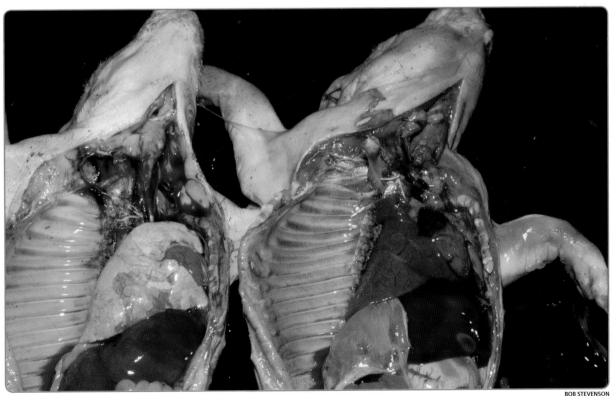

BOB STEVENSON

Post-farrowing care

The sow

Check to see that the afterbirth has been passed and look out for any abnormal discharge. If the afterbirth is retained, the sow may need an injection of oxytocin to get things moving and reduce the risk of infection. Examine the afterbirth for mummified piglets, which may be small and easy to miss. If there are any, plan to vaccinate for porcine parvo virus (see Chapter 5).

Keep an eye on her faeces, as constipation is a common problem in the days after farrowing. Feed plenty of fruit and vegetables to alleviate this, and maybe give a bran mash in place of her usual feed.

Keep your sow on a reduced feed ration until three days after farrowing, then gradually increase by 0.5kg each day, up to a maximum of 7kg a day. Peak milk yield is reached by the third week, and it's important that the sow is eating sufficiently well to cope with the demand placed on her by the piglets.

The piglets

A common procedure in the days following farrowing is the injection of the piglets with iron to protect them against anaemia. Piglets are born with only small reserves of iron; if they were in the wild they'd absorb more each day by rooting in iron-rich soils. When born indoors the only additional iron available is in the mother's milk, which is itself deficient. Piglets are born with normal levels of haemoglobin, a protein found in red blood cells that carries oxygen from the lungs to the cells throughout the body. However, this decreases rapidly, with the result that oxygen is not carried as efficiently around the body and the piglet becomes more at risk of disease. Classic signs of anaemia are the piglets appearing paler from about seven days old, and their skin having a slightly yellow or jaundiced look.

Creep food will provide some iron but piglets won't normally start eating until they're two or three weeks old. The alternative is to give them an iron supplement. This is normally given by injection into a hind leg; or it can be given orally, but can only be absorbed in this way within the first 12 hours of life.

Breeders are split on this one: some say there's no need if the piglets have access to the outdoors regularly, but others swear by a routine jab of iron dextran, regardless of whether they're indoors or out.

Tooth-clipping

Piglets are born with eight sharp 'needle' or 'milk' teeth which can lacerate the sow's teats (and each other) while they compete for available teats, and cause infection. Most sows cope with the irritation, but others may be so uncomfortable that they refuse to feed their litter. Piglets can also do a lot of damage to one another while fighting, and wounds can become infected. For this reason some breeders clip the needle teeth immediately after birth, snapping them off with pliers or grinding them down.

However, the Welfare of Farmed Animals Act (England) Regulations 2007 say this should only be done if absolutely

Below: Injecting iron into a piglet.

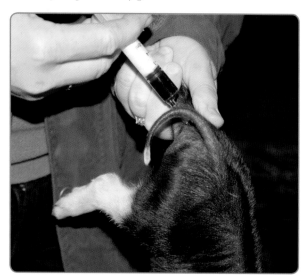

Below: Lacerated teats caused by sharp teeth.

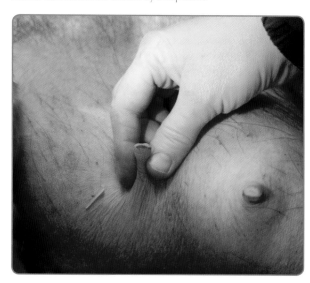

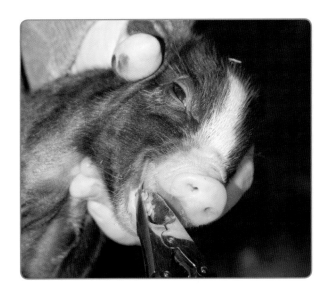

Above: Clipping the teeth of a young piglet.

necessary, where there's evidence of injury to the sow or to other piglets. Grinding is the recommended action, and the procedure must not be carried out after the piglet is seven days old.

Castration

This is a common procedure in commercial systems but not so widespread among smaller producers. Breeders selling pigs as pets will almost certainly – and should – castrate males before sale to prevent unwanted sexual behaviour and aggression. In commercial systems this allows males to be kept together in batches for longer and also removes the risk of boar taint (see Chapter 3).

Castration is regarded as a mutilation under the welfare regulations mentioned above, and guidelines say the procedure can only be carried out 'provided the means employed do not involve tearing of tissue'. The procedure should only be performed by someone competent, and should be done in the first week of life. After the piglet is seven days old it must be carried out by a vet. The guidelines say breeders should consider whether castration is really necessary, and that it should be avoided if possible. Complications can sometimes occur following castration and piglets should be monitored closely.

Tail docking

Again, this is more a feature of large commercial systems where pigs are raised indoors with limited space available, and where tail biting occurs due to stress and boredom. The regulations say this is another form of mutilation, and that it must not be carried out routinely. It can only be done as a last resort once other measures have been tried, eg providing more space, and introducing environmental enrichment devices to keep the pigs entertained. As with castration, it can only be carried out within the first week, after which it must be done by a vet.

The first feed

Colostrum is the rich, first milk produced by a sow and contains essential nutrients and the antibodies piglets need to help prevent disease. It's vitally important that piglets consume adequate colostrum in the first 6 to 12 hours of life, so make sure all piglets get their fair share – particularly late arrivals, which may be weak and are likely to be pushed out of the way by stronger, older piglets.

Cross-fostering

If you have two sows farrowing around the same time and one has a very large litter, it may be possible to 'foster' some piglets on to the other sow. This is best done in the first couple of hours, when the sows are still exhausted from farrowing and sound asleep. Newborn piglets will suckle wherever they can, and once a piglet has started to drink the foster mother's milk and have close contact with those in the natural litter it will start smelling like it belongs to that sow.

Sows that are allowed to farrow together outdoors will frequently have piglets cross-suckling and are more willing to accept it than sows which farrow in a confined space. If you do swap piglets between sows, identify the piglets with stock marker spray so you can tell which belong where.

The next generation

The process of selecting your future breeding stock can begin when the piglets are still with their mother. It's a lot easier checking underlines when the piglets are small and can be lifted easily, but the downside is that you may mistake blind teats for good ones, so you'll have to repeat the exercise when they've grown more and the teats are more prominent.

Before you attempt any kind of inspection, make sure the sow is locked securely out of the pen – preferably out of earshot. This is because when you pick up a piglet it will squeal and the mother's natural reaction is to go charging to the rescue.

Carry out your inspection in a well-lit area with someone to hold each piglet and someone to check the underline and other essential characteristics. This isn't a job you can do alone if you want to get it right. Mark any that you think are good enough for breeding with stock marker spray, and refresh the mark when it starts to fade.

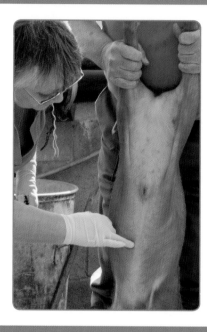

Left: Checking teats on a piglet.

Feeding the piglets

Piglets may start showing an interest in their mother's food from a few days old. A creep feed – small pellets similar in appearance to chicken feed – can be offered from week three. It's best placed in an area where the sow can't reach it, otherwise she'll eat that first before eating her own food. Creep feed is highly palatable and higher in protein than sow pellets – normally between 18 and 25%, depending on the brand. It costs more than everyday pig food but gives piglets a nutritious start.

Gradually the piglets will eat more solid food and take less milk from their mother. Her supply of milk will gradually subside in response to the reduction in demand. As weaning time approaches you can introduce a grower feed, which has slightly less protein. Grower pellets are larger than creep feed, so mix them in gradually until the piglets get used to them. A grower feed can be given right up to pork weight, or you can switch to a specialist 'finisher' feed or sow nuts. Many people who specialise in traditional breeds move gradually from creep feed to sow nuts, because they believe grower feeds – which were designed for modern breeds – aren't necessary for older breeds.

Right: Piglets soon show an interest in solid food.

Weaning

Breeders of traditional pigs tend to wean between eight and ten weeks. In commercial systems, where pigs used are fast-growing modern hybrids, this may be done as early as three weeks, but the Welfare of Farmed Animals Act (England) Regulations 2007 state that it should not be done before the piglets are four weeks old.

You may find that the sow decides to wean the piglets early. Some get fed up with feeding as early as five or six weeks and start shaking the piglets away after they've suckled for a short time. The sows can get restless and bad-tempered and will tell you they want to get out of the pen and back outdoors. If the piglets are sufficiently well grown and healthy and you're happy that they're eating solid food, there's nothing wrong with weaning at this stage.

When weaning day arrives, take the sow away from her litter and place her somewhere out of sight, sound and smell of her offspring. Keep her on the same amount of food for the next three days and then gradually reduce to her normal 'dry' sow amount.

If she's to be put back to the boar, check her body condition and feed her accordingly. If she's lost a great deal of weight during lactation she may need building up to ensure that she's sufficiently well nourished for a good ovulation. In this case you might want to avoid mating her on her next oestrus (which will be in four to seven days' time) and wait another three weeks.

Above and below: Piglets will suckle for as long as the sow allows.

SHOWING

Getting started

If you start breeding pedigree pigs you may find that someone from the breed club or society will encourage you to have a go at showing. Shows are vitally important to clubs in terms of raising awareness of their particular breed, especially if it's one of those on the Rare Breeds Survival Trust Watchlist and needs extra promotion to increase its numbers.

The great benefit for potential pig keepers who may be undecided over what pigs to start off with is that they can see a multitude of different pigs at close quarters and chat to the owners about the pros and cons of keeping them. Many minds are made up at shows and many sales agreed too.

For competitors, shows are great social events. Breeders come from all parts of the UK to compete in the top shows, often travelling hundreds of miles, spending several hours on the road, and sleeping in their trailers for the duration of the event. A lot of the same faces appear at many shows, as keen exhibitors realise the importance of maintaining a presence and showing off their stock. Most people get drawn into exhibiting through someone they know, and so will see at least one friendly face when they arrive. However, it doesn't take long to make friends.

Pig people are extremely helpful and welcoming, and are always pleased to see 'new blood' arriving. The only way to keep pig shows running successfully, year after year, is for new entrants to keep coming in to replace the ones who retire, so it's in everyone's interests to give newcomers a helping hand. The camaraderie doesn't end when the showing is done, either. Pig people love to party, so at the end of a busy day the time is found to relax. Most shows will have a 'stockmen's supper', to which all exhibitors – whatever they're showing – are invited. And if there isn't a sit-down meal organised by the show itself, you can guarantee that the pig people will organise their own event, whether it's a 'bring what you've got' buffet or barbecue, or simply a get-together for a drink in the pig lines.

Below: Showing offers something for all ages.

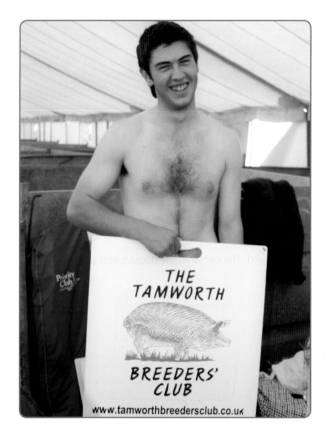

Above: Exhibitors are always game for a laugh once the showing is over.

It doesn't take long to build up a network of friends from far and wide. Some faces you might not see from one year to the next, but each time you meet you'll pick up where you left off last time.

At your first show you'll undoubtedly feel a little like the new kid at school, but it won't be long before some of the other competitors start wandering over for a chat and a look at what you've brought, and you quickly make new friends.

There's no sexism in pig showing. Agricultural shows may have been extremely male-dominated events in the past, but these days there are probably more women than men involved in breeding and showing pigs. There seems to be something about pigs that attracts women more than any other type of livestock, and they do incredibly well with them!

Competition can be a serious business, but it's also great fun. The nice thing about pig breeders is that they don't take themselves too seriously. It may be something to do with the fact that pigs often have such lively and mischievous personalities. You have to expect the unexpected. However well you plan things, a fickle pig with a mind of its own has the ability to turn order into chaos very, very quickly. And the potential for doing so at shows is immense. Fortunately, all breeders have been on the receiving end of piggy disobedience at some time, so rest assured that they'll all rush to your aid with pig boards at the ready should you need help.

A shop window

To the uninitiated, the idea of donning a white coat, picking up a board and stick and parading a pig around a ring in front of a crowd at an agricultural show might seem rather peculiar.

Not everyone likes the idea of showing, but for an increasing number of pedigree pig keepers it's a natural progression from ownership. Shows are shop windows for breeders: many potential customers will travel to shows to see the best of each breed battle it out, and there's no better advertisement for your stock than a fine collection of rosettes pinned up on your pen. A first prize or – even better – a championship rosette is official approval for your herd, and can help to hike up the price of your pigs' future progeny.

The practice of showing livestock goes way back to the 1800s, when agricultural societies started to emerge. Today there are generally two types of show. The biggest and most widespread are those that run through the spring and summer months where competitions concentrate on determining the best of each breed and the best traditional or modern pig. The winter months bring a very different type of event – the pre-Christmas 'fat stock' or 'prime stock' show. Here judges are looking for the best 'finished' stock, the pigs that would catch the eye of a butcher looking for a perfect carcass.

Choosing where to show

Entering a pig in a show isn't like going to Cruft's – you don't have to qualify at a small show in order to win the right to exhibit at a major show. Anyone can enter any show as long as their pigs fit the criteria, so don't be put off going to a big one. At many small village shows pedigrees don't matter and there's often a much more informal atmosphere. consequently some people prefer to start building up confidence locally before travelling further afield. There's no need, however. Showing isn't rocket science, and once you've done one you'll feel much more relaxed about it all. You could even make it a family affair; most shows have young handler categories, so if you have children why not get them involved too?

The first thing you need to do is check which agricultural shows have pig classes. The most prestigious are those sponsored by the British Pig Association (BPA), which are listed on their website at www.britishpigs.org.uk. To enter these you must be a BPA member. Your breed society and your local pig breeders' club or association will also have details of shows.

Many shows have classes for the best sow and litter, and this usually means that you get your pig and her brood into a pen and they get judged there. So if you don't feel confident about jumping in at the deep end and entering the show ring, this might be a good way to get started.

Bear in mind that you can't just turn up on the day at most shows. There are forms to fill in with details of your pig and its pedigree, pig pens to be booked (and human accommodation too, if you're going to be staying over), and, of course, fees to be paid. Show entries normally have to be verified and show catalogues listing details of entries have to be prepared well in advance, so check the closing dates in plenty of time, because you might be surprised at how early they are. A July show, for instance, might have a closing date of the previous April, so do your research early.

You'll also have to think about how movements on and off your holding might affect your plans. Any movement on to your holding triggers a 'standstill' for pigs – which means you can't take any pig off for 20 days, unless it's going for slaughter (see Chapter 2). So, for instance, if you're hiring in a boar, or buying new stock, make sure you allow good time so that the standstill is over before you need to take your existing pigs to the show. The only way around this is to have an approved isolation unit on your holding (see Chapter 5) – a quarantine area where show pigs can be kept separate from the rest of the herd when between shows.

Below: Agricultural shows are always keen to attract new exhibitors – like Stuart Riggs from Abergavenny and his large black.

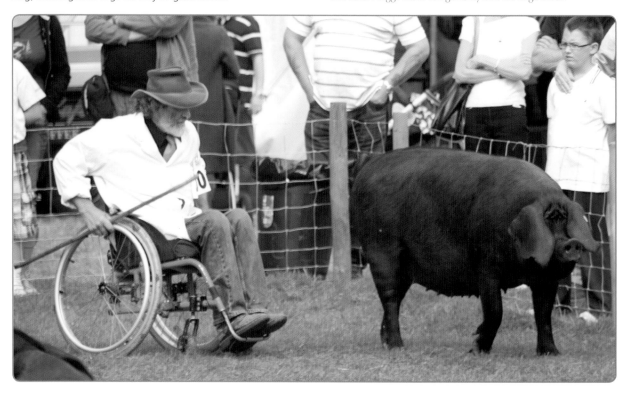

Right: Judge Stephen Hull examines an underline at St Mellans Show, Newport.

Below: A lop and a saddleback being shown at the Hatfield House Country Show.

What can you show?

Experienced pedigree breeders carefully plan their litters to suit the various age classes, so it pays to think ahead. There will normally be classes for young boars, young gilts, sows and senior (mature) boars. The key 'red letter days' in the pig breeding calendar are 1 January, 1 July and 1 September. Entry forms will specify classes for pigs born on or after these dates in the current or a previous year (though not all shows have a September class).

For instance, if you had a gilt born in March this year, it could be shown in the class for gilts born on or after 1 January this year. You have to bear in mind, however, that your gilt will be up against those born in January, which will therefore be older and bigger than yours. Similarly, the 'July class' is for gilts born on or after 1 July last year, and the 'September class' is for gilts born on or after 1 September last year.

Regular exhibitors bear these classifications in mind when planning their litters. They'll aim to have a litter early in the specified month, and work backwards to the day their females need to be served – eg 8 September for a 1 January litter, 8 March for 1 July, or 9 May for 1 September. Having said that, however, it's all down to the judge on the day. A good pig is a good pig.

Birth dates aren't as crucial with older sows; but with young gilts and boars which are still growing, size matters. Sows and gilts that have reached breeding age are deemed to look better

Above and below: Boars need two handlers and only experienced exhibitors should show older boars.

when in pig, so breeders like to put them to the boar a few months before a show. You'll often see in the rules section of the livestock schedule that a pig entered in the sow class must either be expecting a litter or have successfully farrowed and raised a litter in the six months prior to the show. But you should be cautious about transporting a gilt or sow which you believe to be in pig – too soon after serving and the stress of being transported may cause her to miscarry, or prevent successful implantation of eggs, so wait until 21 days after service before you move her anywhere. Similarly, don't take a pig that's too close to farrowing, as the disruption could induce labour. Sows or gilts shouldn't be moved if they're less than three weeks' away from farrowing. Similarly, a sow and litter shouldn't be moved until the piglets are at least three weeks old.

Probably the easiest pig for a newcomer to show is an older sow. They tend to be less excitable and easier to handle. Younger pigs, particularly January-born ones, have boundless energy, are easily distracted, and can have you running around the ring. Young boars can be difficult to handle even when they're just a few months old – which is why shows stipulate that they must be accompanied by two handlers. One takes the lead and the other plays 'backstop', watching that no other boars get too close and blocking any nasty confrontations. Testosterone is a terrible thing in the wrong place, and in the show ring it can be very dangerous indeed. The smell of sex is all around, with females in season within sniffing distance, and male hormones can run riot even in junior boars.

Above: Some shows, like this one at Usk, Monmouthshire, have fun classes, like this pig-racing event.

Senior boars should only ever be taken to shows by extremely experienced handlers, because two mature males can fight to the death – and pose a risk to both exhibitors and spectators – if not properly controlled. Don't assume that because your boar is laid-back and gentle at home he'll be the same at a show. He won't. Most shows specify that adult boars should have their tusks cut before arrival – something which is good practice for all boars, whether they're being shown or not.

Below: The author relaxing after her first show.

Selecting stock

There's often a very relaxed attitude to showing at the smaller agricultural shows, and you sometimes find that villagers are encouraged to enter anything and everything, just to fill the classes and provide entertainment for the visitors. However, at the most prestigious pedigree shows you won't get anywhere unless you show your very best stock.

In reality, the job of selecting stock goes back to before a pig is even conceived. Breeders should be breeding selecting boars and sows/gilts that have the best features and breeding the best to the best.

When it comes to shortlisting your pigs months ahead of a show, the first rule is to go back to the breed standard, the list of essential and desirable points which make a pig eligible for registration (see Chapter 8). If you have pedigree pigs, they should meet the breed standard. If they don't, whoever registered them has sold you something they shouldn't have, or you've registered without paying heed to what your breed society says.

Find your breed society's website or handbook, or check out the relevant section on the BPA's site (www.britishpigs.org.uk) to check the criteria. The list will state all the points that a show judge will be looking for, and each pig will be marked against this list. To recap on the information in Chapter 8, you're looking for

Above: Potential show pigs are often identified months ahead.

Below: An interbreed class at the Royal Welsh Smallholder and Garden Festival.

Above: Laura McLoughlin showing her Berkshire gilt at the Royal Welsh Show.

a good underline, straight, sound legs and a good mouth, plus any characteristics the breed 'wish list' specifies.

If you have a choice of pigs to enter, spend time watching them. See which one catches your eye – the one that has a certain presence and holds him or herself well. Observe them from all angles. Are the backs straight and even, or do they dip behind the shoulders? Watch as the pigs walk: do they walk nice and straight or do they turn their toes inwards or outwards? If in doubt, ask an experienced breeder to take a look for you.

A calm temperament will help in the show ring, because a good-looking pig that's also easy to handle will generally walk better and make life easier for the judge when it comes to observation and examination. It will also make things far more pleasurable for you, the handler. Practice makes perfect, so get the pig used to walking for the show ring well in advance.

Below: A docile pig is an asset in the ring.

Preparation

It should go without saying that your pigs should be in the peak of condition when you show them, and common sense and good husbandry should take care of this. Make sure all your vaccinations are up to date, because shows are to pigs what doctors' surgeries are to humans – places full of nasty things to catch and bring home. Lots of the problems that affect pigs are transmitted when they're in close contact with one another's saliva, faeces or urine, so when you get to the show do what you can to minimise risk.

Check that tags and tattoos are intact. Plastic tags can often get ripped out when your pig scratches its head against something, and tattoos can fade or become illegible. Show officials have been tightening up on identification at shows and checks are made on or soon after arrival to ensure that the pig entered for a particular competition is the one that has been brought to the show. If tags are missing or tattoos cannot be read, a pig may be disqualified.

When you're preparing your pigs for showing bear in mind the same thing as when washing children – remember to clean inside their ears. This will help a lot when the stewards come round checking tattoos with their torches.

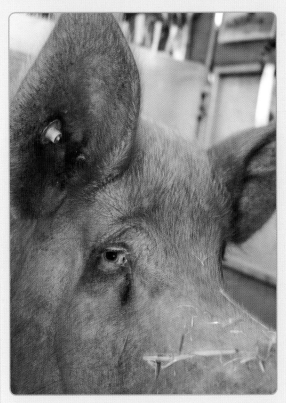

Preparing your pigs

Presentation is important, and it obviously helps your chances if your pigs are clean. This, however, is easier said than done if your pigs live outdoors all year round and have their favourite mud wallows. Some people – particularly those with light-skinned pigs – keep their pigs out of the sun in the months leading up to a show to avoid burning and freckles, and some serious exhibitors will keep their show pigs indoors permanently. Others take their pigs into a barn for preparation a few weeks before an event so that they can be washed and kept as clean as possible. And then there are others who load their pigs straight from the field into the trailer and wash them when they get to the show.

There are always washing facilities at shows, but the standard of the wash pens will vary. Some shows have strong, purpose-built, permanent shower cubicles with hot and cold water on tap; others will have temporary outdoor pens made from galvanised hurdles and you'll have to go and get your own water in a bucket.

Moving nervous pigs – particularly young ones – to the washing area can be a bit of a challenge. Your pens may be a fair old distance away from the washing area, and you may have to negotiate several potential escape routes as well as other people's pigs. It's not unusual to see one or more breaking free and tearing around the rows of pens like a thing possessed.

One alternative is to get them into a barn the day before a show and wash them there to get the worst of the dirt off. Then, when it comes to show time, all they'll need is a quick sprucing-up in the pen. You might even get away with a full

Below: The author gets a pig used to being washed prior to a show.

Gerry Toms

'blanket bath' in the pen – as long as the floor of the pen is concrete. If it's grass you could create a mini mud bath, and your pigs will end up even dirtier than when they arrived.

You have to be comfortable handling your pigs, and bathing them is a good way to help build up a good relationship. Practise at home, long before the show season starts. Particularly when the weather is warm, your pigs will love you if you give them a nice cool rubdown, but remember not to get the water too cold or they may not be grateful.

What to use

Everyone has their own preferences when it comes to pig bathing products. Some use livestock shampoo bought from agricultural merchants, while others choose human shampoos and bubble baths. Both work equally well, but it's worth testing any new products on your pigs well before the show, just in case of skin allergies. Your wash kit should include a selection of brushes, cloths and sponges, plus a few buckets for rinsing and a hook or piece of string so that you can hang your bucket on the side of the wash pen.

All kinds of brushes are available for horses and will do the job equally well for pigs, while a simple nail brush from your bathroom or a kitchen sponge with a scourer on one side will help clean stubborn grime off the back of hocks. It's also worth taking a short length of hosepipe (about 1m to 1.5m should be sufficient), plus a variety of tap connectors and a shower head, just in case there are taps actually inside the pens. This makes the job of washing and rinsing much, much easier.

Some pigs are oiled before going into the show ring, to make their coats shine. Large blacks, Berkshires and British saddlebacks always look extremely smart when sprayed with oil. Specialist pig oil can be purchased, but some people prefer baby oil. Certain breeds – notably Tamworths – cannot be oiled for showing. If oiling is not permitted it should say so in the breed standard. Pig oil can, however, be used in advance of a show to help remove dirt. Massaged into the skin it helps loosen caked-on mud and flakes of dry skin. It can be left on for several days and then washed off just before the show. One word of caution, however: never use oil in sunny weather or if you're going to be showing outdoors on a hot day. The only time you want to smell bacon cooking at a show is at breakfast time!

White-skinned pigs have to look spotless in the ring, but exhibitors are not allowed to use anything to mask unwanted spots or blemishes. It's possible, however, to dust them with wood flour – very fine sawdust, specially sterilised and prepared to be safe with pigs. Exhibitors will wash their pigs thoroughly, then cover their pigs' skin with the powdery shavings. This helps keep them clean and gives the skin a more uniform appearance. A few minutes before going into the show ring most of it is brushed off.

Show ring practice

You should try practising walking your pigs around a makeshift ring well in advance of the show. They need to get used to being guided by pig boards and sticks, and you need to be confident doing it. Make a small enclosure using sheep hurdles – tied together for extra strength – and just have a go. The idea is to 'blinker' the pig's view with the board to help you to move it in the direction you want it to go. On show day you'll also use it to block contact with other exhibitors' pigs. The stick is used to encourage the pig to move along, with taps on the rump and the shoulders as required.

Boards (approximately 60 x 60cm) can be home-made or bought from breed societies and smallholder supplies websites. Occasionally feed suppliers or competition sponsors (banks, for instance) will provide free boards, but the pay-off is that the board will have their logo emblazoned all over it. Nothing is ever entirely free, and you have to decide whether or not you mind being a walking advertisement.

Sticks are normally shepherds' crooks, walking sticks or long, narrow, ruler-like objects with rounded ends, rather like gigantic lollypop sticks. Again, you can make your own or buy one, as there's no standard style. You'll also need a white exhibitor's coat. A lab or catering-type coat will do.

Above: Set up a practice ring at home using hurdles.

You need to keep your left side nearest to the ring, with the pig on your right, and walk in a clockwise direction around the judge. Your stick should be in your right hand, guiding the pig on its right-hand side; the board should be held in your left hand and kept between you and the pig. You can use your board to block other pigs from making contact, but always try not to obscure the judge's view.

Below: Train your pig to walk using a board and stick.

Preparing for the journey

If you haven't used your trailer for a while, check well in advance to see if it's safe and roadworthy. The last thing you want is to have something dreadful happen when your pigs are on board and you're on the way to a show.

Bear in mind, the transportation regulations mentioned in Chapter 2, and make sure you carry with you your certificate of competence. Make sure you set up your eAML2 farm to show movement at least 24 hours in advance, and don't forget to print out a copy of the online movement licence.

What to take with you

- A 'kit box' for all your bits and pieces. This can be a sturdy wooden box or something like a toolbox on wheels.
- First-aid kit for the pigs and for yourself. The show will have its own vets on call should your pigs fall seriously ill, but ask your own vet for advice on what you should take with you.
- Food and drink containers, plus sufficient pig feed for the duration of the show. You may also need to take straw, so check with the show organisers beforehand.
- Washing kit for the final clean-up, along with wellies and waterproof clothing or a big apron.

Below: A well-stocked kit box is essential at any show.

Above: Don't forget your white coat!

- Shovel and brush, for cleaning out the pens each day.
- Mini toolkit including staple gun, hammer, string, cable ties, etc, for putting up an informative display about your herd. You'll also want to put up any rosettes you win!
- Business cards and flyers to advertise your stock.
- Notebook and pens for contact details of potential buyers.
- Your exhibitor's coat, board and stick.
- Small bulldog clips to attach your exhibitor number to the pocket of your coat.
- Food and drink – for you, not the pigs. There can be a lot of hanging around, so it's best to be prepared.

Below: Only rosettes from the actual show you are at should be displayed on your pen.

Arriving at the showground

You will either hand over a copy of the animal movement licence on entry to the showground, or to a steward in the pig section.

When you received confirmation of your entries you should have been sent a plan of the showground and, hopefully, a route to follow to get to the pig pens. If not, stewards will direct you – but be aware that the later you arrive, the busier it will be, so if you're not too confident reversing a trailer plan to get there in plenty of time.

There will be a chief steward for the pig area, who'll tell you where to back up and off-load – but be patient, because he or she might have been there for several hours by the time you get there. Shows depend on volunteers and a generous amount of goodwill, and it's really important to be nice to the stewards. Building up good relationships is crucial, particularly when you're just starting out, so work at it.

You'll be shown where to put your pigs and your kit and, after off-loading, you'll be sent to get your trailer washed out. There will be a designated area for doing this, and there may be staff to do it for you.

If the show is close to home you might be travelling on the day, but often you'll be arriving the night before and camping overnight. You may have decided not to book a tent or caravan pitch, but to sleep in your washed-out trailer instead. Lots of exhibitors do this to cut costs, but remember to take a tarpaulin with you to drape over the roof and sides and cover any vents that might let in the rain.

Below: Make friends with the stewards: they can make your show heaven or hell!

Showtime

One of the most frustrating things about showing pigs – or any animal, for that matter – is finding out what time you'll be in the ring. Catalogues always give a start time, but delays inevitably happen and schedules slip. You have to be prepared for a great deal of hanging around, so don't do a Victor Meldrew or you'll earn a bad reputation before you start. Don't be tempted to nip off to the beer tent, regardless of how bored or nervous you may be.

Keep an ear to the ground and be aware of what's happening in the rings. Make sure you know what time judging starts and make a note of your class number. Everything depends on the number of entries and how thorough the judge is. While you're waiting you can be giving your pigs a final sprucing up.

Exhibitors, too, should always be smartly turned out. Jeans, trainers and wellies are frowned on in the show ring, but still creep in. A shirt and tie worn under a white coat looks neat and businesslike, and demonstrates you're serious about what you're doing. The judge will be dressed in his or her Sunday best, so why shouldn't you?

When your class is called, the stewards should clear the public out of the way so that exhibitors can get their pigs safely into the ring. Efficiency varies incredibly from show to show, so be prepared to weave through people, prams and even dogs at the less safety-conscious shows. Bear in mind too that you won't be the only one moving a pig, so use your pig board and stick to separate your animal from everyone else's, and to minimise nose-to-nose contact with exhibits in the pens you pass.

Once you're safely in the ring give your pig a little time to settle and sniff about before you start trying to take control. Most pigs – particularly younger ones – will want to have a little sprint around. Another thing your pig will want to do when it gets into the ring is to empty itself, so be patient and let nature take its course.

Walking an untrained pig in a clockwise direction around the perimeter of the ring is easier said than done. If your pig wanders off, go with it and do your best to keep it away from the others. Don't run after it – walk briskly and look confident!

It's vitally important to keep an eye on the judge. When he or she comes your way you need to present your pig as best you can. Keep your pig between you

Below: Rachel Nicholas and daughter Catherine with their British lop, Supreme Champion at the Three Counties Show, Malvern.

Above: The author with her sow, Thisbe, who won the title Tamworth Champion of Champions.

Right: Everyone keeps a close eye on the judge – in this case John Millard.

and the judge and try to keep it steady so that it can be examined. A good scratch often does the trick. The judge will want to look at conformation, examine the underline, and see if it meets all the criteria of the breed standard. You'll probably be asked a few questions – age, almost certainly. If you're showing a sow or a gilt the judge may ask whether she's in pig and, if so, when she's due. The more knowledgeable you are about your pig the better. If someone else is showing for you, make sure they're briefed.

Remember to shake the judge's hand and thank him or her if you get a rosette. If you don't win anything, don't be afraid to approach the judge at the end of the class and ask for feedback. It's always useful to know what you could do better next time.

Useful contacts

Breed contacts

Berkshire Pig Breeders' Club
01772 673245
bk.pigs.club@gmail.com
www.berkshirepigs.org.uk

**British Kunekune
Pig Society**
01799 525421
secretary@britishkunekunesociety.org.uk
www.britishkunekunesociety.org.uk

British landrace
02879 386287
info@deerpark-pigs.com
www.deerpark-pigs.com

British Lop Pig Society
01948 880243/07759 487469
secretary@britishloppig.org.uk
www.britishloppig.org.uk

**British Saddleback
Breeders' Club**
07967 019332
admin@saddlebacks.org.uk
www.saddlebacks.org.uk

Duroc
01244 851705
andrew@wtms.biz

**Gloucestershire Old Spots
Breeders' Club**
mail@oldspots.org.uk
www.gospbc.co.uk

Hampshire
01767 650884
balshampigs@btinternet.com

**Large Black Pig
Breeders' Club**
jwood@largeblackpigs.co.uk
www.largeblackpigs.co.uk

Large white
01954 719263 or 07966 582804
portbredypigs@hotmail.co.uk

Mangalitza
02879 386287
info@deerpark-pigs.com
www.deerpark-pigs.com

Middle white
07980 542067 or 01772 673245
gracebank.pigs@yahoo.co.uk

Oxford Sandy and Black Pig Society
01993 881207
info@fieldfarmoxford.co.uk
www.oxfordsandypigs.co.uk

Piétrain
01293 711529
gillofawr@talktalk.net
www.pedigree-pietrain-pigs.co.uk

Tamworth Breeders' Club
07747 034170
secretary@tamworthbreedersclub.co.uk
www.tamworthbreedersclub.co.uk

The Pedigree Welsh Pig Society
info@pedigreewelsh.com
www.pedigreewelsh.com

Useful contacts for getting started – including getting CPH and herd numbers

England: Rural Payments Agency
0300 200 301
enquiries@rpa.gsi.gov.uk
www.rpa.gov.uk

**Northern Ireland:
Department of Agriculture,
Environment, and Rural Affairs**
0300 200 7852
email daerahelpline@daera-ni.gov.uk.
www.daera-ni.gov.uk

**Republic of Ireland:
Department of Agriculture, Food,
and the Marine**
01 607 2000
info@agriculture.gov.ie
www.agriculture.gov.ie

**Scotland:
Rural Payments and Services**
0300 244 4000
www.ruralpayments.org

Wales: Rural Payments Wales
0300 062 5004
rpwonline@wales.gsi.gov.uk
www.wales.gov.uk/rpwonline

Other useful sources
Animal and Plant Health Agency
0300 200 301
apha.corporatecorrespondence@
    apha.gsi.gov.uk
www.gov.uk/government/organisations/
    animal-and-plant-health-agency

**Artificial insemination
– pedigree semen suppliers:
Deerpark Pedigree Pigs**
02879 386287
info@deerpark-pigs.com
www.deerpark-pigs.com

British Pig Association (BPA)
01223 845100
bpa@britishpigs.org
www.britishpigs.org.uk

British Veterinary Association,
020 7636 6541
bvahq@bva.co.uk
www.bva.co.uk

**Department for Environment,
Food, and Rural Affairs (DEFRA)**
08459 335577
helpline@defra.gsi.gov.uk
www.defra.gov.uk

Food Standards Agency
helpline@foodstandards.gsi.gov.uk
020 7276 8829
www.food.gov.uk

Humane Slaughter Association
01582 831919
info@hsa.org.uk
www.hsa.org.uk

**National Farmers' Retail
& Markets Association (FARMA)**
0845 4588420
info@farma.org.uk
www.farma.org.uk

Rare Breeds Survival Trust,
Stoneleigh
024 7669 6551
www.rbst.org.uk

Soil Association
0117 314 5000
www.soil-association.org

Magazines

Country Smallholding (monthly)
01392 888481
www.countrysmallholding.com

Home Farmer (monthly)
0845 2260477
www.homefarmer.co.uk

Pig World (monthly)
01430 810597
www.pigworld.org

Practical Pigs (quarterly)
01959 541444
www.kelseyshop.co.uk

Smallholder (monthly)
01354 741538
www.smallholder.co.uk

Books

Beautiful Pigs
(Frances Lincoln, 2009) Case, Andy

Cured
(Jacqui Small, 2010) Wildsmith, Lindy

Garth Stockmanship Standards
(5M Enterprises Ltd, 1998) Carr, John

*Higgledy Piggledy:
The Ultimate Pig Miscellany*
(Quiller, 2010) Lutwyche, Richard

*Managing Pig Health
and the Treatment of Disease*
(5M Enterprises Ltd, 1997) Muirhead,
Michael R. and Alexander, Thomas J.L.

Manual of a Traditional Bacon Curer
(Merlin Unwin, 2009) Davies, Maynard

*Pig Ailments – Recognition
and Treatment*
(Crowood Press, 2005) White, Mark

Pig Signals: Look, Think, Act
(Context Products, 2006)
Hulsen, Jan and Scheepens, Kees

*Recognising and Treating
Pig Diseases*
(5M Enterprises Ltd, 1998) Muirhead,
Michael R. and Alexander, Thomas J.L.

*Recognising and Treating
Pig Infertility*
(5M Enterprises Ltd, 1998) Muirhead,
Michael R. and Alexander, Thomas J.L.

Sausage
(Lyons Press, 1998) Livingstone, A.D.

The Commuter Pig Keeper
(5M Publishing, 2016) Giles, Michaela

The Sausage Book
(Farming Books, 2006), Peacock, Paul

*Sow: A Practical Guide to Lactation
and Fertility Management*
(Context Products, 2007) Scheepens,
Kees and van Engen, Marrit

*The Whole Hog: Exploring the
Extraordinary Potential of Pigs*
(Profile Books, 2005) Watson, Lyall

Websites and discussion forums

Pig Progress
www.pigprogress.net

The Accidental Smallholder
www.accidentalsmallholder.net

The Pig Site
www.thepigsite.com

Glossary

Ad lib feeding – Allowing free access to food at all times.

Ark, arc – An outdoor pig house, normally made of a wood frame and galvanised metal sheets.

Artificial insemination (AI) – Method of inserting semen into a pig using a catheter instead of mating her with a boar.

Back fat – Layer of fat along the pig's back used to gauge overall carcass condition.

Bacon – Cured pork.

Bacon pig, baconer – A pig being reared for bacon rather than pork, and which will be slaughtered between 80kg and 100kg.

Boar – An uncastrated male pig.

Boar taint – An unpleasant odour or flavour in meat from some, but not all, mature male pigs.

BPA – British Pig Association.

BPEX – British Pig Executive.

Carcass – The body of a pig following slaughter and removal of internal organs.

Castrate – See *Hog*.

Colostrum – The first milk produced after farrowing, rich in nutrients and antibodies.

Conformation – The thickness of muscle and fat in relation to the size of an animal's skeleton, *ie* the 'shape' of the carcass profile and degree of muscularity.

CPH number – County Parish Holding number.

Creep area – A section of the farrowing pen separated off for piglets.

Creep feed – Very small, high-protein pellets specially formulated for piglets.

Dam – The female parent of an animal.

Dead weight – The weight of a pig carcass once slaughtered and eviscerated.

Dished – Used to describe the face of a pig with an upturned snout, *eg* Tamworth or Berkshire.

Dressing – Preparing the carcass ready for butchering.

Dry sow – One which is not in pig (pregnant).

Entire – Uncastrated and therefore capable of reproduction.

Farrowing – The process of giving birth.

FCI – Food chain information.

Fecundity – In relation to a sow, her reproductive success, or ability to produce healthy piglets.

Finished pig – A pig ready for slaughter.

FSA – Food Standards Agency.

Gestation period – Length of pregnancy, which is around three months, three weeks and three days.

Gilt – A female pig that hasn't yet produced a litter.

Ham – Cured pork, normally from the back legs.

Hog – More common in the US. A castrated male pig, also known as a 'castrate'.

Hogging – In female pigs, showing signs of being in season.

In pig – Pregnant.

Isolation unit – A dedicated place, approved by the Animal Health, where pigs can be 'quarantined' to avoid the statutory 'standstill' movement restrictions, eg for pigs travelling to and from shows, stud boars, etc.

Killing-out percentage (KO%) – The weight of the eviscerated or 'dressed' carcass as a proportion of the live weight of the animal prior to slaughter.

Lactation – The period during which a sow produces milk.

Lairage – An area in an abattoir where animals are held before slaughter.

Line breeding – Process of breeding closely related animals to continue certain favoured characteristics.

Litter – Collective name for piglets born to a sow.

Live weight – Total weight of pig before slaughter.

Maiden gilt – A sexually mature pig that hasn't yet been mated.

Marbling – Visible deposits of intramuscular fat (fat found within the muscles).

Mummy, mummified piglet – A foetus which dies in the womb before maturing and appears as a dark brown, incomplete piglet at birth.

Pedigree pig – A pure-bred animal which has been officially registered.

Piglet – Young pig.

Pork – Fresh, uncured pig meat.

Porker – A pig reared to pork weight (normally around 60kg), slaughtered at around five to seven months old.

Progeny – Offspring.

Prolificacy – The productiveness of a sow, in terms of numbers of healthy piglets per litter.

RBST – Rare Breeds Survival Trust.

Scours, scouring – Diarrhoea.

Sire – The male parent of an animal (noun); to father offspring (verb).

Slap mark – An identification stamp imprinted on the back of a pig before slaughter.

Sow – A female pig after her first litter.

Store – A pig being fattened for meat.

Suckling pig – A young (often two to three weeks old) piglet that's taken from its mother to be cooked whole. Considered a delicacy in many parts of the world.

Swill – Waste human food which it is now illegal to feed to pigs.

Swine – Term used by Americans to describe pigs.

Taint – See *Boar taint*.

Terminal sire – In cross-breeding, a boar chosen specifically to pass on particular traits.

Underline – The double row of nipples in male and female pigs.

Weaner – A piglet separated from its mother and eating solid food. Weaning normally takes place between five and ten weeks.

Withdrawal period – In veterinary medicines, the time during which an animal cannot go into the food chain following treatment.

Acknowledgements

I am indebted to Bob Stevenson from Usk Vets, for all the help and advice given in preparing the health and breeding sections of this book. Bob is the consultant vet to the British Pig Association, a past president of the British Veterinary Association, and a man who is amazingly generous with his time and always willing to share his knowledge and experience. At a time when there are comparatively few farm vets around, and even fewer with specialist knowledge of pigs, I have been very fortunate in being able to call upon his expertise. There is little that Bob doesn't know about pigs and I was delighted when he agreed to check the health and breeding sections of the book content for accuracy, as well as supplying many of the gory images!

Rachel Nicholas, who owns the Dukes Herd of Oxford sandy & blacks, deserves a special mention for supplying the piglet which graces the cover. She helped out in an entertainingly chaotic photo-shoot which gave the intrepid and patient photographer James Davies the most challenging assignment of his career. I am also grateful to Mary Riggs for giving me a lesson in AI and allowing me to photograph her inseminating her large blacks. And. thanks to her tuition, I am now the proud owner of my first AI litter.

Finally, many thanks to the publishing team – particularly Project Manager Louise McIntyre, who had the confidence to hire me as an author, but also to all those who had to deal with my bizarre requests for anatomical illustrations. I am still amazed some of those emails made it through the spam filter.

Index